Climate Change and Global Warming... Exposed

Hidden Evidence, Disguised Plans

Andrew Johnson
August 2017

Text Copyright Andrew Johnson © 2017.

Cover by Neil Geddes-Ward

Please analyse, discuss and share.

"I wonder if we could contrive ... some magnificent myth that would in itself carry conviction to our whole community."

Plato

Table of Contents

Acknowledgements iv
Introduction 1
Climate Change? Or is it Global Warming? 6
Framing the Issue, Controlling Thought and Debate 10
A Brief History of "Global Warming" and Attitudes to It 16
What Fundamentalist AGW Beliefs Might Result in 32
"Don't talk about Changes in the Solar System!" 48
Persistent Jet Trails/Chemtrails 68
Studying Contrails/PJT 98
Are Contrail Patterns Caused by Commercial Airline Flights? 107
Perception Management of the PJT/Chemtrail Issue 138
Weather Anomalies 150
Geoengineering – Civilian and Military 176
Wilhelm Reich, Cloud Busting and Orgone Energy 187
"Don't Talk about This Weather …" 200
What Fundamentalist AGW Beliefs ARE Resulting in …. 216
An Answer to Environmental Pollution And Destruction? 234
Conclusions 240
References 247

ACKNOWLEDGEMENTS

My thanks go out to all these people (in no particular order) for their support, for sharing their research and for giving me encouragement in one way or another:

Dr Judy Wood
Richard D Hall
Daniel Elliot
Sarah Johnson
Chris Johnson
Anthony Beckett
Sally Kennedy

Phil Morris
Menna Trinder-Widdess
Debbee Tapner
Neil Geddes-Ward
Brendan Houlihan
Adam Dwyer

Thanks to **Mohan Rao** and **Carl James** for proof reading and corrections. Also, enormous thanks are due to **Chris Goj** for his thorough proof-reading and very helpful and practical suggestions for improving this text.

About The Author

Andrew Johnson grew up in Yorkshire, England and graduated from Lancaster University in 1986 with a degree in Computer Science and Physics. He spent most of the next 20 years working in Software Engineering and Software Development and travelled to several different countries. He has also worked full and part time in lecturing and tutoring (in Adult Education). Now he works for the Open University (part time) tutoring students, whilst occasionally working freelance on various small software development projects. In 2006, he set up the website www.checktheevidence.com as a repository of information he had been studying. Andrew is married and has two children.

INTRODUCTION

"You are never dedicated to do something you have complete confidence in. No one is fanatically shouting that the sun is going to rise tomorrow. They *know* it's going to rise tomorrow. When people are fanatically dedicated to political or religious faiths or any other kind of dogmas or goals, it's always because these dogmas or goals are in doubt."

Robert T. Pirsig, *Zen and the Art of Motorcycle Maintenance*, Pt. II, Ch. 13

Why this Book Exists

In about 2003, I realized I was being lied to – about enormously important topics. You can read about this on my website if you so wish, though it may change your view of the validity of what I present here. This book exists to expose people to new information and challenge them to be honest about it. By the end, I hope that I have shattered your world view and you will, like me, see things in a completely different light and begin to understand how multiple parties seem to have conspired to trick us and make us go along with draconian plans. My main aim is to expose some of the lies and tactics on which these plans for humanity are based. Of course, now that I have used language like this, you may have little or no desire to read any further. You may now consider that anything I write, from hereon in, must be biased and unreasonable – "the ramblings of a paranoid delusional crank/conspiracy theorist…" That's fine by me – it's your loss, not mine. I know things that you don't. Goodbye …

Still reading? Good – I thank you for your kind attention! The evidence shown in this book affects our immediate and long-term future in extremely powerful ways – yet most people are unaware of it. Or, if they are aware of the evidence, they want to ignore it. I would argue that the ignorance of this evidence has been carefully engineered – using all manner of psychological tactics and perception management techniques (which we will also discuss).

Have You Heard …?

I think it is probably fair to say that most of the people that need to read this book probably won't ever hear about it. If they did hear about it, I don't think it is very likely they would read it. I hope that I am wrong about my opening statements, but only if I am right about the conclusions I draw at the end of this book.

The weather is something which affects all of us, every day of our lives. It dictates our long and short-term plans. In the UK, the weather is the "default topic" of conversation for strangers and friends alike. Though it has obvious and far reaching importance, most people don't know very much about the weather. These days, they don't take time to observe it and probably spend more time looking at a screen than they do looking at the sky. They don't really know why the weather seems to be changing. They are told by mainstream sources that anthropogenic (human) activity is having a negative effect on the weather. Well, when I say "told," I really mean that the idea is "rammed down their throats at almost every opportunity and if they disagree with what they are being force-fed they must be mad – or stupid."

I can be fairly confident in saying that this book collects together, for the first time anywhere, a range of diverse data which proves that the whole issue of "climate change" is more complicated and challenging than *almost all* researchers are willing to consider, examine, or entertain. For example, this book contains astronomical data which most climatologists will not discuss in full. Similarly, the book contains climate and weather data that astronomers will not discuss. The book contains some data that neither astronomers nor climatologists will discuss. It contains some data that no scientists will appropriately discuss.

Certain parts of the evidence in this book are never accurately or appropriately discussed by mainstream academic or media outlets. However, there are also important parts of this book that are never discussed by those *challenging* the mainstream views either. That is, significant parts of the evidence in this book is rarely discussed or even mentioned on popular "alternative knowledge" or "conspiracy" type websites.

If you are a regular web-surfer you can use the evidence I bring forth in this book to further your own research. Others will likely find many reasons why you should just ignore what I have written. There are hundreds or thousands of websites devoted to pointing out errors in thinking of the so-called errant ways of "global warming sceptics" and "climate change deniers" (whatever that means).

So, if you are a sincere and honest person, you might want to actually check that what I present to you is true. I must state that I am neither a climatologist nor a meteorologist. I do have a degree in Computer Science and Physics and I do work (part time) for a university, though this book has absolutely no connection to the work I do for that university.

My formal education included a study of physics and chemistry and some aspects of weather and climate, when I studied O Level Geography. (For example, I know what an Occlusion is). Additionally, I have studied Astronomy for quite a number of years on a casual basis. In any case, some of the aerial features here documented are so blatantly anomalous (primarily due to their geometry) that I would strongly argue that only basic observation skills are needed to see that something real and disturbing (but not officially acknowledged) is occurring.

I have never had any scientific peer-reviewed papers published in any scientific journals – but some of my work has been read and reviewed by one or more scientists. So, if you want to rely on author credentials alone in relation to what is, after all, a hugely important issue, you can conclude now that I can't possibly have anything relevant to say, can I? So, there's no point in you reading much further …

Wait a minute – may I ask you …. "Do you need a meteorologist to tell you when it is raining?" That is, do you still have basic powers of observation and reasoning that allow you to evaluate what is happening in the world, without needing an expert to confirm or refute your own conclusions …? Do you need permission from an academic, an expert or someone in authority to know whether something is true or not? Of course, only you can answer these questions yourself. I answered these questions for myself long ago.

Corrections and Augmentations? Yes, Please!

I take the stance of a witness being cross-examined in a court case, and so I say to you dear reader, "I promise to tell the truth, the whole truth and nothing but the truth, so help me web!" Hence, *when you have read through the whole book and checked all the references*, if you find any errors or inappropriate references, please contact me and tell me what they are, so that I may correct and improve this work – so that we all benefit. I have attempted to reference this material to a good standard, but of course, it is not being "peer reviewed" for publication in a recognized scientific journal or even a magazine. I doubt any journal would publish the evidence I include and describe here – it would be too damaging to established interests. Indeed, if there is a knowledgeable reader, who has had papers published in climatology or meteorology – or any related – types of journals, I challenge them to take some of what I have presented here, work it up into a paper suitable for submission, and attempt to get it published. Please write to me and tell me of your experience.

When I was writing the book, one suggestion I had was to include a blank page at the end of each chapter for readers to make notes on. I have included a space for some notes at the start of each chapter.

Beliefs and Knowledge

This book is primarily about knowledge, not belief. It is not about believing whether climate change has one particular cause or another, it is about showing you evidence that you probably have never heard about, and it should lead you to a conclusion. If you believe and have faith in government think-tanks and academic institutions and you believe they can act without any self-interest and you believe they always tell the "whole truth" – and you believe they always admit when they are wrong and then correct their errors, again, there is little point in you reading any further. Going through this book, you will probably find the earlier sections less challenging than the later ones, though in the next chapter, you may find yourself a bit out of your depth and feeling uncomfortable with what I am saying. You have been warned.

Global Warming – "Cause and Effect"

I am certainly not convinced that the average global temperatures are indeed rising and I am also *not convinced,* even if average temperatures are rising, that this will lead us to catastrophe. I *am convinced* that those that claim the cause of any warming is solely the result of human activity are not being honest. Hence, they are trying to solve an *imaginary problem*. They claim to know the cause of disputed data/observations about global temperatures and therefore they claim to be able to develop solutions. The truth is, they don't know what's happening and they don't really know whether they are dealing with a cause or an effect. An example here is the observation that due to the timing or time period associated with most of the data, a cogent argument can be made that CO_2 increase in the atmosphere is an *effect* of global warming, *not a cause*. Many sites which promote the Anthropogenic Global Warming (AGW) theory, such as "Real Climate Science" [1] claim they have explained this, whilst others say the CO_2 increase follows warming.[2]

What you will see towards the end of this book is evidence of other climate or weather effects that *all* the main researchers and scientists are ignoring. Decades-long arguments over the temporal relationship between changes in global temperature and CO_2 levels cause the ignorance of much more easily observable effects. Hence, scientists who are "pro" or "con" AGW theory will not solve the problem they are setting out to solve – *because both "pro" and "con" groups are wrong!* Any action taken, based on a false conclusion, could make the problem worse – and one has to wonder, is that what someone else wants.

CHAPTER ONE
CLIMATE CHANGE? OR IS IT GLOBAL WARMING?

"All great truths begin as blasphemies."

George Bernard Shaw

Space for Notes Below

"Global Warming" became "Climate Change …"
As some people have observed, we were originally presented with the idea of "Global Warming" but many legitimate questions were raised about this description, and so at some point the language morphed into "Climate Change." This is actually much harder to be critical of – because it is such a vague description. Everyone should know that the climate is *always* changing. The underlying problem is identifying the reason for this. Similarly, if weather forecasters are reluctant to try and forecast weather more than about three days into the future, how is it possible to determine what changes will happen years from now? Is it true then that longer term models can be more accurate than short term ones? Or is it true that, as with financial investments, "Past performance is no guarantee of future performance"?

Distinguishing between "Environmental Damage" and "Climate Change"
Before anyone accuses me of not being worried about my own "environmental footprint" I will state that I myself, like so many others, attempt to use all resources responsibly and not waste anything – even down to making sure that paper is printed/written on both sides. I use sites like "freecycle" and ebay to ensure that manufactured goods stay out of landfill sites for as long as possible.

There is no doubt that human activity is damaging the environment. For example, industrial pollution and "industrialized" fishing and agricultural practices have, it is clear, destroyed habitats and caused the extinction of a large number of species of flora and fauna – in various ways, and for various reasons. With the real damage caused on a daily (and large-scale) basis, it is easy to manipulate people's emotions, fears and "environmental conscience" when presenting these matters (without proper distinction between "Environmental Damage" and "Climate Change").

Most people have at least some awareness of the effects of their lifestyle on the environment, though these effects are often hidden – except when large numbers of people are affected by such things as dumping of industrial waste, depletion of fish stocks or intensive farming practices etc. Is it any wonder that some websites promote the AGW myth "out of love"[3]?

A distinction must also be made between local and global climate changes. For example, the Urban Heat Island Effect[4] is well understood and the data is clear. The same conclusions (by definition), cannot apply on a global scale, however because an "island" is not the same as "a globe"!

Nowadays, when some people suggest that changes in the weather are *not* caused by industrial pollution, it is assumed they are somehow suggesting we should *not* worry about wholesale pollution of the environment. Or it is tacitly assumed that those suggesting other causes for changes in climate are saying that "humanity should not worry about the effects of their activities on the environment." Therefore, because of the increase in visibility of environmental organizations and issues in the last 30-40 years, it becomes easier to marginalize the legitimate scientific questions raised by people who remain unconvinced that carbon dioxide emissions (specifically) from human industrial activity have had any provable effect on global climate in the last 150 years.

There is little doubt that CO_2 output has increased due to industrialization but how that increase is measured, relative to all other sources of this simple gas, is a far more difficult question (which too many people have claimed to know the answer to). Later in this book, you will read about much more data *which needs to be properly studied*. This data proves the climate change issue is much more complicated than is generally spoken of – and extra CO_2 from industrialisation is probably the least significant issue.

I won't go through all the enormous problems with the AGW CO_2 theory – others, such as Dr Tim Ball and Dr Don Easterbrook, have done a more than adequate job of this – and that is not the primary focus of this book. However, I do cover some of the more salient points which show that the sooner this theory is filed in the "Not-enough-supporting-evidence" category, the better it will be for everyone.

The Difference Between Modelling and Observation

Most of the theories about AGW (Anthropogenic Global Warming) involve some kind of modelling or projection – this is always a less reliable basis for a proposal than "observed data from recent history" (which is largely what the conclusions in this book hinge on). One question that can be raised, for example, is "How is the distinction made between the quantity of naturally produced CO_2 and that produced by industrial processes?" My understanding is that such figures must be based on estimates – as it is impossible to accurately quantify these things on a global scale.) Again, the argument about these details is not the focus of this book – there are many alternative books you can read about that.

"Saving the Planet" – Underlying Assumptions

Since the late 1980s or early 1990s, the debate about what is most affecting our climate has been raging. More recently, it seems, certain groups within mainstream science and academia have gained the upper hand and the wider population appears to have accepted their (flawed) conclusions. These conclusions have then been used to set public policy in governments around the world. This has happened despite the repeated reports of a failure to agree various "environmental targets" at a number of "climate summits" that have taken place over the last few years (Kyoto[5], Copenhagen[6] and so on). As I will show you later, a troubling development in the last few years has been open discussion of geoengineering – modifying the earth's climate directly to ameliorate the effects of the supposedly human-induced climate problems.

For completeness, I will now state the underlying problems which are associated with the unproven Anthropogenic Global Warming (AGW) theory:

- Human activity is producing "greenhouse gases" (we hear incessantly about carbon dioxide and "carbon footprints")
- There are too many consumers.
- Consumption is damaging the environment.
- Steps need to be taken to "change habits" and stop or reduce environmental damage.
- We humans need to live in a "sustainable way."

Or, to say this another way "It's YOUR fault, ye know! *You* consume the goods! *You* drive the cars! *You* use the electricity! Global Warming is entirely YOUR FAULT! And you complain at the idea of being taxed for your *'Carbon Footprint'!* Some people!"

Of course, the ideas of carbon footprints, carbon credits etc have been written into government policies and EU directives, even though there is no firm evidence that anthropogenic CO_2 has affected global temperature in any significant way. One among many of the fundamental issues is the true source of any problematic CO_2. A volcanic eruption, for example, produces vast quantities of this "greenhouse gas" and there is little to nothing a government think tank, university or climatologist can do about that. One of the other key issues seems to be that, according to ice core and other physical/geological records, we have had much warmer temperatures on earth than now – before we had any anthropogenic CO_2. Whilst I will spend some time on these issues in this book – the AGW/non-AGW arguments are not meant to be a focus of this book – I just want to point out some of the issues and that people do argue over them. I also want to point out how one side has more executive power and authority than the other, and how that power is wielded.

CHAPTER TWO
FRAMING THE ISSUE, CONTROLLING THOUGHT AND DEBATE

"Facts do not cease to exist because they are ignored."

Aldous Huxley

Space for Notes Below

A Warning from History

In the research I have done, it has become clear that corporate and even military interests in matters such as climate cannot be overlooked or ignored. For example, the climate *is* a matter of national security. Therefore, perhaps readers should bear in mind President Eisenhower's message, from his 1961 farewell address to America[7]:

> *"In the counsels of Government, we must guard against the acquisition of unwarranted influence, whether sought or unsought, by the Military Industrial Complex. The potential for the disastrous rise of misplaced power exists, and will persist. We must never let the weight of this combination endanger our liberties or democratic processes. We should take nothing for granted. Only an alert and knowledgeable citizenry can compel the proper meshing of the huge industrial and military machinery of defense with our peaceful methods and goals so that security and liberty may prosper together."*

In this same speech, in relation to scientific research, he also said:

> *...the free university, historically the fountainhead of free ideas and scientific discovery, has experienced a revolution in the conduct of research. Partly because of the huge costs involved, a government contract becomes virtually a substitute for intellectual curiosity. For every old blackboard, there are now hundreds of new electronic computers. The prospect of domination of the nation's scholars by Federal employment, project allocations, and the power of money is ever present – and is gravely to be regarded. Yet, in holding scientific research and discovery in respect, as we should, we must also be alert to the equal and opposite danger **that public policy could itself become the captive of a scientific-technological elite.***

But why would the military industrial complex be interested in climate change? Just as importantly, isn't it now true that we are controlled in relation to climate issues by just such a "a scientific-technological elite." How would this "elite" group make "public policy" its captive, do you think? Remember, this came from a former US President – long before global warming was ever considered as something real ...

Hegelian Dialectic – "Divide and Conquer" and "The Control Group"

It seems to be true that if an authority can get ordinary people to believe their way of life is under threat and it presents them with a vaguely plausible story as to how they know the threat is real, the people that believe the story will become easier to manipulate. The ordinary people may (occasionally) question the authority about its conclusions, to check they are being told the truth. If said authority can answer their questions in a credible way, it will gain their confidence and, as a consequence, gain power over them.

The "credible story" that authority gives may have several layers and these layers mask the underlying agenda of the authority concerned. Authority can then use various techniques to "frame" any debate and this is generally much easier when emotive accounts can be drawn on – such as when people, particularly children, are injured or killed. Ordinary people may not even perceive when the authority is actually manipulating them – or that there is in fact, a "control group" that is steering things…

If you want to get your own way with an issue – pass laws, make changes, there is a technique by which this is more easily achieved. You have parties on each side of the issue and then you get them to argue about the issue. All the while, the arguing parties distract most people from seeing additional related evidence and issues – because observers are too busy watching the "infighting" and/or siding with one party or another. As Richard D Hall would say, they have all fallen for a "phoney bone of contention."

This is essentially another technique, which can be used as a kind of "backdrop" to something called the Delphi Technique. The Delphi Technique[8] is used to encourage people to come to a particular, desirable conclusion about something, so that one or more parts of a larger plan can be put in place, without people noticing what that actual plan is. One way of describing such manipulation is as follows[9]:

> *The Hegelian dialectic is the framework for guiding our thoughts and actions into conflicts that lead us to a predetermined solution. If we do not understand how the Hegelian dialectic shapes our perceptions of the world, then we do not know how we are helping to implement the vision. When we remain locked into dialectical thinking, we cannot see out of the box.*

To add to this description:

> *The synthetic Hegelian solution to all these conflicts can't be introduced unless we all take a side that will advance the agenda. The Marxist's global agenda is moving along at breakneck speed. The only way to completely stop the privacy invasions, expanding domestic police powers, land grabs, insane wars against inanimate objects (and transient verbs), covert actions, and outright assaults on individual liberty, is to step outside the dialectic. This releases us from the limitations of controlled and guided thought.*
>
> *Hegelian conflicts steer every political arena on the planet, from the United Nations to the major American political parties, all the way down to local school boards and community councils. Dialogues and consensus-building are primary tools of the dialectic, and terror and intimidation are also acceptable formats for obtaining the goal. The ultimate Third Way agenda is world government. Once we get what's really going on, we can cut the strings and move our lives in original directions outside the confines of the dialectical madness.*

Others have succinctly described this tactic as consisting of three elements – "problem-reaction-solution." Another version, used in marketing, is even shorter "hurt and rescue," where you point out to someone they must have a real or potential problem that they weren't even aware of and then you offer to "sell them the solution."

At this point, I put it to you that to really understand what is happening in climate science, one has to entertain that there is, an overall "control group." The control group isn't a single country, nor a single company – it is some other entity. This group has an agenda – which has nothing to do with "saving the planet and protecting the environment." The agenda is one of gaining increased power and control over the lives of "ordinary people." By the end of this book, this reality should be easier to see, but identification of the control group will not be especially clear.

In climate science (and many other areas of science it seems), research and campaign groups are polarised. It then becomes much easier to hide the control group's identity and even hide the very idea of the control group's existence! The task of concealment is therefore assisted by a process of creating "factions" within the general populace who "sit" on either side of an issue and expend their energies "arguing" and even attacking each other. Whilst attention is focused on the squabbles, the control group can carry out its agenda – whilst most people are blind to the evidence which betrays the control group's reality. (The clearest evidence of the reality of some kind of powerful control group – of unknown identity – is shown in the data from Chapter Twelve and likely some of the effects observed in the data in Chapter Six are a result of this control group's activities.)

Most people are also blind to the operation of the control group's tactics to hide their own existence and identity, and the control group's ability to confuse or confound anyone trying to uncover more information about it. Some people become "seduced" by the wish for more power and influence and then, perhaps, are unwittingly co-opted into the larger agenda. Some people feel they are helping the earth / planet / environment or they are helping to create a "positive future."

Asch Conformity Experiment

This experiment clearly shows how easy it is to generate a consensus – in a person who will agree with a conclusion even though it is blatantly false[10]. In our case, the consensus being built is that AGW is real – and this is the "side" that is almost exclusively promoted by the mainstream media.

A simple experiment using actors and an unwitting "victim"[11] shows them a simple diagram with vertical lines to compare, and state which line is the same length as another one. In the diagram below line C is the same length as the example. The "victim" must indicate which line he thinks is the same length as the example (correct answer C). The actors deliberately give incorrect answers.

> Asch measured the number of times each participant conformed to the majority view. On average, about one third (32%) of the participants who were placed in this situation went along and conformed with the clearly incorrect majority on the critical trials.
>
> Over the 12 critical trials about 75% of participants conformed at least once and 25% of participants never conformed. In the control group, with no pressure to conform to confederates, less than 1% of participants gave the wrong answer.
>
> Conclusion: Why did the participants conform so readily? When they were interviewed after the experiment, most of them said that they did not really believe their conforming answers, but had gone along with the group for fear of being ridiculed or thought "peculiar." A few of them said that they really did believe the group's answers were correct.

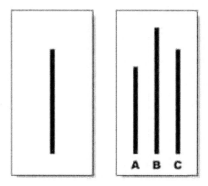

Figure 1 – Asch Conformity experiment – example diagram.

It is important to note that this was an isolated experimental run. Participants were *not* subjected to repetition of "A, A, A" (using the diagram above where the wrong answer of "line A is the same length as the example" is rammed down people's throats) multiple times per day/week etc, as is the case with what is shown in hourly news bulletins; i.e. in the Asch experiment, the "wrong conclusion" was not programmatically repeated for weeks, months and years to influence the "victim" – it was only exposed to them a few times.

I will later show you what appears to be an example of widespread usage of subliminal cues to acclimatise you to an atmospheric phenomenon, so that you won't even notice it – let alone ask questions about it (as I have).

We will talk more later about "creating consensus" – and whether this has been done in relation to AGW. You should already know that we have a dialectic in the form of AGW/non-AGW groups.

CHAPTER THREE
A BRIEF HISTORY OF "GLOBAL WARMING" AND ATTITUDES TO IT

"It is one thing to show a man that he is in an error, and another to put him in possession of truth."

John Locke (1632–1704) *An Essay Concerning Human Understanding*, Bk. IV, Ch. 7

Space for Notes Below

A Theory and a Scam
To the person who studies all the available evidence openly and honestly (i.e. they go far beyond talk of atmospheric CO_2 and water vapour levels), it should become clear that the AGW theory is a global scam (I will, henceforth be referring to it as a scam as much as a theory). The scam gained popularity since approximately the end of the 1980s. In the 1970s, the thinking was that a new Ice Age was about to begin. Since the early 1990s, a relentless campaign of propaganda has been in operation, employing vast resources to cajole and convince the general population that we must "change our errant ways", to reduce our "carbon footprint" (one of the most meaningless phrases ever coined) or face extinction. This campaign has employed many so-called scientists and even Hollywood stars and other artists[12] and it has been a huge success. Perhaps because people aren't able to distinguish between environmental destruction and pollution, environmentalists can manipulate them – almost at will. The bottom line is that most people "do the best with what they are given."

It is this suppression, I would argue, that has therefore enabled the "hidden power structure" to control people's lives – because they claim that the standard technologies and processes in use are too destructive, dirty or dangerous, and so we all need to be told what we can and can't do.

Global Warming and the Club of Rome.
It appears that the basis of the global warming scam may have been "dreamt up" by a think tank called "The Club of Rome" – which "advises" the United Nations (more about them later) on various issues.

Indeed a synopsis of their 1991 work "The First Global Revolution: A Report by the Council of The Club of Rome" reads[13]:

> *A blueprint for the 21st century, this report by the Club of Rome sets forth a strategy for world survival at the onset of the first global revolution on a small planet we seem hell-bent to destroy. With conclusions that are even more far-reaching that those of its Limits to Growth report 20 years ago, this new global revolution comes into being amid social, economic, technological, and cultural upheavals, and calls for an all-out attack on world crises.*

Even more surprising, perhaps, is an often-quoted passage on page 75 of this report:

> **The common enemy of humanity is Man**
>
> *In searching for a common enemy against whom we can unite, we came up with the idea that pollution, the threat of global warming, water shortages, famine and the like, would fit the bill. In their totality and their interactions these phenomena do constitute a common threat which must be confronted by everyone together. But in designating these dangers as the enemy, we fall into the trap, which we have already warned readers about, namely mistaking symptoms for causes. All these dangers are caused by human intervention in natural processes, and it is only through changed attitudes and behaviour that they can be overcome. The real enemy then, is humanity itself.*

So, it seems that this "club" decided on the tactics they could use to manipulate the global population. Of course, once the propaganda had been in place for decades, who would ever believe that this passage represents the original intent of these people? You can read plenty of statements which claim to debunk the idea that I present above. However, I say again, *look at all the evidence in this book* before making a final judgement.

We now consider how Pope Francis himself (also associated with Rome, of course!) has fuelled the propaganda/lie machine as recently as September 2015:

He said[14]:

> *"Mr President, I find it encouraging that you are proposing an initiative for reducing air pollution. Accepting the urgency, it seems clear to me Climate change is a problem which can no longer be left to a future generation."*

So, either the Pope himself has knowingly become a propagandist, or he has just not studied the available evidence which shows how bogus the "air pollution" story is.

And of course, it's not only the Club of Rome that pushes the propaganda – it is pushed by both left and right wing political parties, NGOs and most if not all environmental groups like "Friends of the Earth" and so on.

Who's Who in Global Warming

One of the main figures who was responsible for starting the pattern of thought that there was such a thing as anthropogenic (human-made) global warming was James Lovelock. His invention of the electron capture detector in 1957 gave scientists a way of more accurately measuring levels of atmospheric pollution – such as CFCs (chlorofluorocarbons). Later, he theorised that the Earth operates as a single, living organism and reacts to those other life forms damaging or infecting it. This was called the Gaia theory (and was a big part of the plot line in a 1985 BBC Thriller series called "Edge of Darkness"[15])

However, in April 2012, even Lovelock said in an interview to msnbc.com[16] that he had been unduly 'alarmist' about climate change.

A small note is that Lovelock, in 1954, did receive some funding from the Rockefeller Travelling Fellowship in Medicine[17]. (Rockefeller is a name that seems to come up quite frequently when studying environmentalism.)

Figure 2 – Maurice Strong (deceased) and Kofi Annan

Another name which comes up in researching the beginnings of the global warming scam is Maurice Strong who was a Canadian multimillionaire. In 1947, he started work as a clerk at the UN Offices in New York, where he first met none other than David Rockefeller. By the 1960s, Maurice Strong had become wealthy from Canada's oil industry. According to an article in the Daily Telegraph[18] ...

> *Strong came to see that the key to his vision was 'environmentalism', the one cause the UN could harness to make itself a **truly powerful world government**.*

In an article on Maurice Strong's own website by Ruud Lubbers, former Prime Minister of the Netherlands, Lubbers writes[19] (emphasis added):

> Indeed, China represents the need and possibility to create a truly multi-polar world. The challenge for climate change and sustainable development is very much about China. Maurice Strong, Mikhail Gorbachev, **Steven Rockefeller** and myself, we invested enormously in the Earth Charter Initiative

It should come as no surprise that it was essentially Strong who was responsible for setting up the 1992 Rio Summit[20] – where something called "Agenda 21" was first revealed. It is essentially a set of "guidelines" referencing all walks of life – and even education … We will cover Agenda 21 in more detail later.

Attitudes to Those Challenging Some of the Science

Typically, when data like that shown later in this book, is brought up or discussed, there is a huge adverse reaction to it, and a "big fight" then ensues. Here is a typical example of this – in a TV interview with Ed Begley and Stuart Varney of Fox News, where Varney asks if the science about climate change is conclusive and challenges various statements made by Begley, Begley states "You're spewing your nonsense again …"[21]

Similarly, famous people – who are not scientists – have made widely read comments which do not address any of the data which is included later in this book – e.g. "[Paul] *McCartney, in Interview, Compares Global Warming Sceptics to Holocaust Deniers*"[22]. For those who are naturally sceptical, sniggers or cries of phrases such as "Conspiracy Theory" or "Tin Foil Hat" often serve to discourage renewed, reasoned and dispassionate analysis of evidence. This, it seems, is an example of how group mentality can be influenced and moulded.

It should be borne in mind that finding the truth demands a most vigorous application of energy and time to investigate what is truly happening – and this investigation must take place with a clear and stated independence from corporate and other vested interests – and without recourse to either invective or preconceived notions.

Doublethink

This was a concept that was made famous in George Orwell's *1984*, where ordinary people could be made to have thoughts which contradicted each other without ever considering that such thoughts were indeed contradictory. It seems that this has now happened with some of those who wish to promote the belief in anthropogenic global warming.

In a posting by Dr Richard Parncutt in October 2012, he implied we should consider that people who "didn't agree with climate change" should be put to death[23].

> Let us assume for the purpose of argument that the global warming skeptics are right with a probability of 90%. That is a very generous assumption, if we consider that hundreds of highly qualified, internationally leading scientists have devoted their lives to this problem, and in general disagree. According to this assumption, the scientists are right with a probability of 10%. Let us also assume that 10 million lives are at risk as a result of global warming in coming decades, which is also a very conservative estimate. In this case, questioning the existence or human origin of global warming – in a way that slows down current attempts to stop it – is directly risking the lives of one million people (10% of 10 million).

So, this person wants to protect the planet from the effects of global warming but wants to consider the death penalty for those that provide evidence that it's not happening in the way he (and others who have a passionately held belief in the AGW or GW theory) says it is. Parncutt therefore seems to be advocating the killing of humans to save human lives – based on a theory (and he agrees there is uncertainty). He didn't even realize his contradiction – that's doublethink – especially as he was a member of another organization that ignores evidence of enormous human rights abuses, Amnesty International! Unsurprisingly, this academic had to retract these remarks[24].

> I wish to apologize publicly to all those who were offended by texts that were previously posted at this address. I made claims that were incorrect and comparisons that were completely inappropriate, which I deeply regret. I alone am entirely responsible for the content of those texts, which I hereby withdraw in their entirety. I would also like to thank all those who took the time and trouble to share their thoughts in emails.

But then still he said:

> The following extract from the text was intended to apply to the entire text: "Please note that I am not directly suggesting that the threat of execution be carried out. I am simply presenting a logical argument. I am neither a politician nor a lawyer. I am just thinking aloud about an important problem."

Well, perhaps people like him need to take a long hard look at how media programming and exclusion of coverage of important evidence can lead to extremist beliefs, even in high-minded academics.

Critics of AGW – Exposing the Scam is Bad for Your Career ...

Here I document several examples of harsh treatment of those who have challenged the AGW propagandists. For example, it was only after he retired that long-time newsreader Peter Sissons criticised the BBC for failing to be more sceptical about AGW[25]. Chris Landsea (resigned from IPCC) has now admitted some data they used came from a mountaineering magazine article and a student's dissertation[26].

Figure 3 – Dr David Bellamy

It seems to be true that those who haven't been affected by "programming", "education" or indoctrination are "dealt with" by those that *have* been affected... (this is very similar to what happens with the Thought Police in Orwell's *1984*).

Dr David Bellamy, a hugely popular science presenter on the BBC at one time, said he "didn't get any more phone calls from the BBC" when he started to point out the flaws in the AGW science[27]. He is quoted[28] as saying about global warming, "This is not science – it's religion." His scepticism meant the end of his career as he had known it. He said that "They froze me out, because I don't believe in global warming. My career dried up. I was thrown out of my own conservation groups and I got spat at in London."

Another popular BBC science presenter, Johnny Ball, also rejects any consensus and was reported as saying[29]:

> *"In the past decade or so I've been mocked, vilified, besmirched – I've even been booed off a theatre stage – simply for expressing the*

> view that the case for global warming and climate change, and in particular the emphasis on the damage caused by carbon dioxide, the so-called greenhouse gas that is going to do for us all, has been massively over-stated.

A further example of the work of the climate change "thought police" [30] was on the case of Quentin Letts who hosted a BBC 4 series called "What's the Point Of." This light-hearted series poked fun at various organisations. One particular episode was about the UK Met Office, and the subject of climate change came up. Letts is reported as saying:

> "I was accused of having shown disrespect to climate change. Mr Lilley had cracked a joke: 'They [the Met Office] come before the Select Committee on Energy and Climate Change ... and tell us they need even more money for even bigger computers so they can be even more precisely wrong in future.' I chuckled. I had 'not reflected prevailing scientific opinion' about global warming. "Letts asked: "Er, hang on, chaps. No one ever told me that. Why on earth would independent journalists accept such a stricture? Why should climate change be given such special protection?"

This episode of the programme was deleted from iPlayer prematurely[31].

A further example was seen when Philippe Verdier, weather chief at France Télévisions[32], the country's state broadcaster, was "sent on a forced holiday" for publishing a book questioning the causes of any climate change. He did this following a "meeting with Laurent Fabius, the French foreign minister, who summoned the country's main weather presenters and urged them to mention 'climate chaos' in their forecasts."

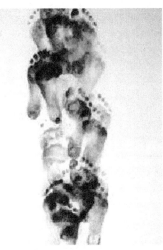

"I walked through some charcoal
and I couldn't believe that I had left
so many carbon footprints"!!!
I felt dirty and shamed!

Nevertheless, there are still a sizable number of people who can see the AGW belief system for what it is. However, few or none of them have ever discussed some of the evidence I present later in this book. Neither do they seem to understand how they themselves have been manipulated into a role of, pardon the pun, "clouding the issue ..."

The Myth of Climate Change/Global Warming and a '97%' Consensus on the Causes

Can we draw a comparison to the "Emperor's New Clothes" fable? I think we can. Frequently quoted is a figure that "97% of scientists agree on the causes of climate change." What is the origin of this constantly repeated false belief[33]? We can read in the *Wall Street Journal* ...

> *This figure comes from a survey "The "97 percent" figure in the Zimmerman/Doran survey represents the views of only 79 respondents who listed climate science as an area of expertise and said they published more than half of their recent peer-reviewed papers on climate change. Seventy-nine scientists—of the 3,146 who responded to the survey – does not a consensus make."*

Also see a presentation by Dr Don Easterbrook[34] for some detailed information which dismantles the arguments put forward by those promoting the anthropic-caused climate change/global warming myth.

A study of available information shows that many scientists don't go along with the AGW scam/theory. For example, less than half of all published scientists endorse it:

> *Of 528 total papers on climate change, only 38 (7%) gave an explicit endorsement of the consensus. If one considers "implicit" endorsement (accepting the consensus without explicit statement), the figure rises to 45%. However, while only 32 papers (6%) reject the consensus outright, the largest category (48%) are neutral papers, refusing to either accept or reject the hypothesis. This is no "consensus."*[35]

We can also see that 100 scientists told Obama he is wrong.

> *"Few challenges facing America and the world are more urgent than combating climate change. The science is beyond dispute and the facts are clear."*
> — PRESIDENT-ELECT BARACK OBAMA, NOVEMBER 19, 2008
>
> *With all due respect Mr. President, that is not true.*

> *We, the undersigned scientists, maintain that the case for alarm regarding climate change is grossly overstated. Surface temperature changes over the past century have been episodic and modest and there has been no net global warming for over a decade now.*[1,2] *After controlling for population growth and property values, there has been no increase in damages from severe weather-related events.*[3] *The computer models forecasting rapid temperature change abjectly fail to explain recent climate behaviour.*[4] *Mr. President, your characterization of the scientific facts regarding climate change and the degree of certainty informing the scientific debate is simply incorrect.*[36]

This statement by President Barack Obama serves to illustrate the power of the control group – and the results it can achieve when there is a general and even wilful ignorance and suppression of additional and contrary evidence in relation to the issue at hand.

Other vociferous sceptics are Dr Tim Ball (University of Winnipeg)[37] and Dr Harold Lewis (Emeritus Professor of Physics, University of California). Lewis said "global warming is the greatest and most successful pseudoscientific fraud

I have seen in my long life" and "I think it is the money, exactly what Eisenhower warned about a half-century ago. There are indeed trillions of dollars involved, to say nothing of the fame and glory (and frequent trips to exotic islands) that go with being a member of the club." [38] In the UK, Lord Christopher Monckton has continued to speak out against the scam, though his record appears somewhat chequered [39].

Another articulate speaker, who accurately characterises much of the talk of AGW as "propaganda", is Dr Richard Lindzen[40] – though AGW theorists sully him with connections to Exxon Mobil. [41] How many AGW theorists, however, have stopped driving cars, stopped using oil based products and stopped buying or using plastics excessively …? (We will get to the whole oil and fossil fuel scam later.)

The Global Warming Petition Project was set up by Arthur B. Robinson, BS Caltech, PhD UCSD and Noah E. Robinson, BS SOU, PhD Caltech[42], and is circulated with a summary of peer-reviewed research[43]. The petition has already been signed by over 30,000 scientists.

> *We urge the United States government to reject the global warming agreement that was written in Kyoto, Japan in December, 1997, and any other similar proposals. The proposed limits on greenhouse gases would harm the environment, hinder the advance of science and technology, and damage the health and welfare of mankind.*
>
> *There is no convincing scientific evidence that human release of carbon dioxide, methane, or other greenhouse gases is causing or will, in the foreseeable future, cause catastrophic heating of the Earth's atmosphere and disruption of the Earth's climate. Moreover, there is substantial scientific evidence that increases in atmospheric carbon dioxide produce many beneficial effects upon the natural plant and animal environments of the Earth.*

An article by Jens Biscof called *Ice In The Greenhouse: Earth May Be Cooling, Not Warming* has an important paragraph, reproduced below. The last sentence of said paragraph is especially important.

> *Indeed, there are signs from some natural systems that global warming is under way. Observations of the pack-ice thickness of the Arctic Ocean from submarines with upward-looking sonar, for example, show a thinning trend since the 1970s. The margin of permafrost is moving north, and the vegetation in the high*

> northern parts of the world is changing toward more temperate forms. **But it is by no means clear whether these signs indicate real, worrying proof of manmade, permanent and potentially disastrous climate change, or just regular, naturally occurring variations in the Earth's climate system.**[44]

Biscof is author of *Ice Drift, Ocean Circulation and Climate Change* and is a research assistant professor in Old Dominion's Department of Ocean, Earth and Atmospheric Sciences.

But, it now seems we have a propagandised consensus[45], which claims "the debate is over." However, when a debate ignores contradictory and pertinent evidence (as given in the links above), debate is not over because any conclusions from that debate (that have already been implemented in government policy) are provably wrong. It is clear to me "the con is in place" – there is an unwritten law "don't question it … or else!" As we have seen, those disagreeing with consensus on climate change have been ostracised – the "thought police" have been busy since the early 1990s. The idea of warming *has to be global*. It seems that people won't be happy that any warming might be just localised.

Light Relief

Having covered this, I could not resist including this cartoon[46] (see Chapter Five to understand what the cartoon refers to).

Figure 4 – "Man created global warming in seven days…"

Fossil Fuel – Is it Really from Fossils?

Few people stop to consider, more in the case of crude oil than in the case of coal and similar fuels, whether the "fossil" moniker is appropriate. In simple terms, some of the oil is extracted from depths far below where fossilized remains have ever been found. An interesting page by Col L Fletcher Prouty contains a quote from an August 2002 article/paper, published in the Proceedings of the National Academy of Sciences (US), which had a partial title of "The genesis of hydrocarbons and the origin of petroleum."[47] Dr. Kenney and three Russian co-authors conclude:

> The Hydrogen-Carbon system does not spontaneously evolve hydrocarbons at pressures less than 30 Kbar, even in the most favorable environment. The H-C system evolves hydrocarbons under pressures found in the mantle of the Earth and at temperatures consistent with that environment.

In a video interview Prouty contends that in 1892, the Rockefeller family (more on them later) influenced[48] attendees of The Geneva Congress on Organic Nomenclature[49] to conclude that crude oil must be composed of formerly living (fossilised) material – because it consisted mainly of organic compounds. It must therefore be a "fossil fuel" – which could "run out" at any time. Therefore, it could become a scarce (and therefore more valuable) resource. It seems the label stuck and nowadays, people remain confused about the differences between the chemical and biological meanings of the word "organic." In chemistry, organic compounds contain carbon (and usually hydrogen too). In biology, "organic" typically means "living" or perhaps material which is made by a living organism.

(Also, we have further confusion, as "organic" is now used to describe horticultural and agricultural cultivation methods. Just about all the food we eat is "organic" – whether it has had pesticides sprayed onto it or not!)

That some oil wells have refilled themselves does lend support to the conclusion that the oil itself is continually being created by processes in the earth's mantle.

Again, I do not include this information because I think that it means we don't need to worry about the amount of natural resources we use and burn – rather, I include it to point out yet another related area where we appear to have been lied to.

The Peak Oil Scam and "Fossil Fuels"

In the early 2000s, there was renewed talk of us reaching "peak oil" – indeed, even former Cabinet Minister and would-be labour leader Michael Meacher had an article published in the UK *Guardian* newspaper where he expressed scepticism of the official account of 9/11[50] and expounded the view that it was a US plot to invade Iraq and other countries in the region to take control of oil supplies. (Look what has happened since …) Around the same time, other authors and researchers were arguing about whether we had reached peak oil production. Here we are, in 2017 and no one is talking about this issue (fuel prices are currently lower than they were 10 years ago).

"Climategate" – Admitted Science Fraud

This did receive some small media coverage in 2010, but in comparison to the significance of the data revealed, it was not covered properly. The whole set of e-mails "leaked" just before the Copenhagen Summit in 2009, can be downloaded[51] and I contend that the timing of the "leak" is part of the general Hegelian Dialectic ("Divide and Conquer" strategy) already described.

Even though the fraud here is overt and obvious, no one was prosecuted, no government or EU policies changed and, as far as I know, no funding was withdrawn from those who committed fraud.

Here are some examples:

Manipulation of evidence:

> *I've just completed Mike's Nature trick of adding in the real temps to each series for the last 20 years (ie from 1981 onwards) and from 1961 for Keith's to hide the decline.*

Notice the use of the words "trick" and "hide" – this seems to betray an agenda which is not one of establishing the truth. An agenda of honest inquiry would foster a comment like "we should re-check our models – maybe they are not correct" rather then "how can we massage the evidence so that our models still look good?"

Private doubts about whether the world really is heating up:

> *The fact is that we can't account for the lack of warming at the moment and it is a travesty that we can't. The CERES data published in the August BAMS 09 supplement on 2008 shows there should be even more warming: but the data are surely wrong. Our observing system is inadequate.*

Suppression of evidence:

> *Can you delete any emails you may have had with Keith re AR4? Keith will do likewise. He's not in at the moment – minor family crisis. Can you also email Gene and get him to do the same? I don't*

> have his new email address. We will be getting Caspar to do likewise.

Fantasies of violence against prominent Climate Sceptic scientists:

> Next time I see Pat Michaels at a scientific meeting, I'll be tempted to beat the crap out of him. Very tempted.

Attempts to disguise the inconvenient truth of the Medieval Warm Period (MWP):

> ... Phil and I have recently submitted a paper using about a dozen NH records that fit this category, and many of which are available nearly 2K back–I think that trying to adopt a timeframe of 2K, rather than the usual 1K, addresses a good earlier point that Peck made w/ regard to the memo, that it would be nice to try to "contain" the putative "MWP", even if we don't yet have a hemispheric mean reconstruction available that far back ....

"Climategate 2"

Further problems appeared for the AGW believers over the so called "Hockey Stick" graph which allegedly shows sharp increases in global temperatures in the last few decades (but as with all other measures, cannot be used to conclusively prove the cause of any alleged rises – which is the real issue.) In reading up on the issue, it is not clear whether some sites are promoting the AGW scam or disagreeing with it. Indeed, one site discussing the "hockey stick"[52] graph has the tag line "Getting Skeptical about Global Warming Skepticism." Wow. However, the title of one book seems to tally with the conclusion I have come to – *The Hockey Stick Illusion; Climategate and the Corruption of Science (Independent Minds)*[53]

You will then go on to find articles about law suits and FOIA's being issued. None of this, of course, helps us establish what is causing the "climate problem" – let alone what the true nature of that problem itself is.

Conclusion

Perhaps this whole chapter proves the theory that ...

> "To every PhD, there is an equal and opposite PhD."

Unfortunately, one set of PhDs helps to set government policies and endorses ever increasing regulation of people's daily lives (more on that later …) – all based on unproven, incomplete and propagandized science.

CHAPTER FOUR
WHAT FUNDAMENTALIST AGW BELIEFS MIGHT RESULT IN ...

"Science ... commits suicide when it adopts a creed."

T. H. Huxley (1825–95) *Darwiniana*, 'The Darwin Memorial'

Space for Notes Below

"Let's Geoengineer!"

After decades of AGW/global warming propaganda, promoted by various individuals and groups, including non-academic unelected bodies in the UN, we begin to see the proposals for schemes to solve this imaginary problem.

To me, it seems the word "geoengineering" has slowly been introduced into the public's vocabulary since approximately the start of the new millennium. I have a 1996 *Chambers CD ROM English Dictionary* and it does not include this word. The word can be defined as "a means of changing the earth's weather or climate by artificial methods." However, the earliest use of the word that I can find is in 1961, according to Google's NGRAM viewer. [54]

Since about 2008, there has been open discussion[55,56,57,58] of "hypothetical" geoengineering techniques such as "SRM – Solar Radiation Management" to "block out the sun." In 2008, suddenly, we in the UK had a Department of Energy and Climate Change (DECC). They were quick to get "geo-engineering" into their vocabulary. Their short report published in April 2009 – titled *Geo-engineering Options for Mitigating Climate Change*[59] – mentions nothing about aircraft trails though it mentions the potential use of Stratospheric and Tropospheric aerosols in SRM.

Commenting on his study of geoengineering published in Autumn 2009, Royal Society Member Prof John Shepherd said,[60]

> "None of the geoengineering technologies so far suggested is a magic bullet; all have risks and uncertainties associated with them."

A proposal by Dr David Keith involves the atmospheric distribution of aerosol compounds, such as sulphur or "micro discs" of aluminium oxide and barium titanate.[61]

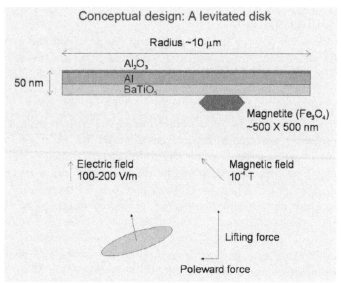

Figure 5 – David Keith's levitated reflective nano disk.

Keith was quite conspicuous at the Royal Society's public consultation meeting I attended in November 2010. Titled "Geoengineering – Taking Control of our Planet's Climate" (discussed later), the idea of this meeting fitted with a statement on page 3 of the UK Government's 2010 document "The Regulation of Geoengineering"[62], thus:

> "... the UK and other governments need to push geoengineering up the international agenda and get processes moving."

House of Commons
Science and Technology
Committee

The Regulation of Geoengineering

Fifth Report of Session 2009–10

Report, together with formal minutes, oral and written evidence

Ordered by the House of Commons to be printed 10 March 2010

Figure 6 - UK Parliament – "Regulation of Geoengineering" – 10 March 2010 [63]

Reading further on page 3 of the report, we can again observe the effects of 20+ years of propaganda and thought-policing. The results are rather disturbing:

> *There are three reasons why, we believe, regulation is needed. First, in the future some geoengineering techniques may allow a single country unilaterally to affect the climate. Second, some—albeit very small scale—geoengineering testing is already underway. Third, we may need geoengineering as a "Plan B" if, in the event of the failure of "Plan A"—the reduction of greenhouse gases—we are faced with highly disruptive climate change. If we start work now it will provide the opportunity to explore fully the technological, environmental, political and regulatory issues.*

In September 2010, the Government Response[64] to the earlier *"Regulation of Geoengineering"* report recommended that

> *"the UK Government press the governments of other countries to adopt a similar approach to SRM research."*

It also states that a ban on SRM testing would be

> *"unenforceable and be counter-productive."*

Announced in September 2011, the now "stalled" SPICE project, led by a consortium of UK Universities and Marshall Aerospace[65] proposed introducing "reflective aerosol particles" into the atmosphere. But in February 2012, Matt Andersson, a former aerospace and defence advisor said in a letter to the UK *Guardian* newspaper[66] that

> *Few in the civil sector fully understand that geoengineering is primarily a military science and has nothing to do with either cooling the planet or lowering carbon emissions.*

I agree – as I have now come to the inescapable conclusion that secret geoengineering has been conducted since at least 2001 (discussed in Chapter Thirteen).

In a paper titled "The Geoengineering Option – A Last Resort Against Global Warming?"[67] (Stanford University), we see someone's idea of solving the "problems" created by AGW.

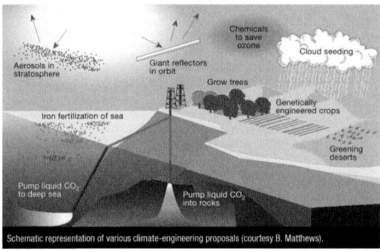

Figure 7 – "The Geoengineering Option" (Are they serious? [68])

So, the final plan would use:
- aerosols in the stratosphere
- giant reflectors in orbit
- iron fertilization of the sea
- genetically engineered crops
- greening deserts
- cloud seeding
- chemicals to save ozone ...

This all stems from this kind of thinking, outlined in their paper "A Last Resort ...", mentioned above, in which they write:

> *The world's slow progress in cutting carbon emissions and the looming danger that the climate could take a sudden turn for the worse, require policymakers to take a closer look at emergency strategies for curbing the effects of global warming.*

In short, "we're all gonna die, so we need to do something really desperate... now!"

Legislation and Government Reports Mentioning Geoengineering and Weather Modification

Here, I present a sample of what is "out there" – and you can probably find more. The situation currently, as I understand it, is that in the "white/open world" (as opposed the "black/secret world"), there are only research proposals and there are large-scale projects that are acknowledged to be in operation at the time of writing this book (2017).

Congress Bill S. 517: Weather Modification R & D

In 2005, Senator Kay Bailey Hutchison put this bill before the 109th US Congress[69]:

SEC. 2. PURPOSE.

> *It is the purpose of this Act to develop and implement a comprehensive and coordinated national weather modification policy and a national cooperative Federal and State program of weather modification research and development.*

The original posting on the website looked like this:

109th U.S. Congress (2005-2006)
S. 517: Weather Modification Research and Development Policy Authorization Act of 2005

| Overview | ⇨ Summary | ⇨ Floor Speeches | ⇨ Other Info |

Official Title: A bill to establish a Weather Modification Operations and Research Board, and for other purposes.

Status:	**Scheduled for Debate**
	This bill was considered in committee which has recommended it be considered by the Senate as a whole. Although it has been placed on a calendar of business, the order in which bills are considered and voted on is determined by the majority party leadership.
Introduced:	Mar 3, 2005
Last Action:	Dec 8, 2005: Placed on Senate Legislative Calendar under General Orders. Calendar No. 319.
Sponsor:	Sen. Kay Hutchison [R-TX] (no cosponsors)

Figure 8 – S.517 Weather Modification Bill

UN Resolution – Environmental Modification (1976)

Few people are aware that the subject of Environmental Modification was brought up in a UN Resolution 31/72. In a document dated 10 December 1976[70], the resolution reads:

> *Convention on the Prohibition of Military or Any Other Hostile Use of Environmental Modification Techniques*
>
> *Adopted by Resolution 31/72 of the United Nations General Assembly on 10 December 1976. The Convention was opened for signature at Geneva on 18 May 1977.*
>
> *ARTICLE I*
> *1. Each State Party to this Convention undertakes not to engage in military or any other hostile use of environmental modification techniques having widespread, long-lasting or severe effects as the means of destruction, damage or injury to any other State Party.*
>
> *2. Each State Party to this Convention undertakes not to assist, encourage or induce any State, group of States or international organization to engage in activities contrary to the provisions of paragraph 1 of this article.*

USA – Congress Bill H. R. 2977 – Proposing a Ban on Space Based Weapons.

As shown below, this bill[71] was put before congress on 2nd October 2001 by Denis Kucinich. The version shown is the first draft of the bill, but it was later revised and some terminology was removed.

Figure 9 – Denis Kucinich (he's a bit older now!)

107TH CONGRESS
1ST SESSION **H. R. 2977**

To preserve the cooperative, peaceful uses of space for the benefit of all humankind by permanently prohibiting the basing of weapons in space by the United States, and to require the President to take action to adopt and implement a world treaty banning space-based weapons.

IN THE HOUSE OF REPRESENTATIVES

OCTOBER 2, 2001

Mr. KUCINICH introduced the following bill; which was referred to the Committee on Science, and in addition to the Committees on Armed Services, and International Relations, for a period to be subsequently determined by the Speaker, in each case for consideration of such provisions as fall within the jurisdiction of the committee concerned

Figure 10 – HR 2977

On page 5 of this bill, it defines some of the types of weapons which should be included in the ban:

> *(B) Such terms include exotic weapons systems such as—*
> *(i) electronic, psychotronic, or information weapons;*
> *(ii) chemtrails;*
> *(iii) high altitude ultra low frequency weapons systems;*
> *(iv) plasma, electromagnetic, sonic, or ultrasonic weapons;*
> *(v) laser weapons systems;*
> *(vi) strategic, theater, tactical, or extra-terrestrial weapons; and*
> *(vii) chemical, biological, environmental, climate, or tectonic weapons.*

What *was* Mr Kucinich thinking?

Royal Society Meeting – 2010 – Geoengineering – taking control of our planet's climate

I attended two days of these public conferences/consultation meetings – one day in 2010 and one day in 2013. The venue was the Royal Society's Wellcome Trust Lecture theatre (as in the Wellcome Pharmaceutical Company).

Figure 11 - 2010 – "Geoengineering – taking control of our planet's climate." Is it only me that is scared by the title? [72]

All of the presentations assumed that CO_2 was the "evil gas which needed to be stopped" (hardly stopping to consider it is the gas we all breathe out.) Below, I have included a further discussion of the event and the lectures/talks that were given.

It was in response to the nonsense discussed at this meeting that I originally started to compile what has now become this book. I submitted my report to a body called SRMGI – the Solar Radiation Management Governance Initiative. It contained a summarised version of some of the (pre-2011) data which is in this book. Of course, my submission was completely ignored, as I found out when a similar meeting took place in 2013.

There were perhaps 150-200 attendees seemingly mainly from universities and some businesses with a few lay people like myself. There were only three people in the audience who spoke up to challenge those "wearing the scientific blinkers." One member of the audience referred to "denialists" towards the end of the day with a comment to the effect that their conclusions were "disproportionately represented" in the press (it is not clear to me where he got that idea from).

Climatologist David Keith seemed to be there to compliment everyone and try to augment what presenters had said. To me, it felt like he was guiding the event somehow – as if he was kind of "top of the tree." To give you a flavour of the presentations, I have included descriptions below:

An Overview of Proposed Enhanced Weathering Methods

Dr Tim Kruger of the University of Oxford (Martin School)

This paper looked at the various methods for changing the balance of CO_2 by various "absorption" techniques. There were some surprising suggestions such as spreading olivine powder, dumping limestone into the ocean (in regions of upwelling current). Kruger reported that he had been investigating a technique proposed by Kheshgi in 1995 through his company Cquestrate[73]. In the Q & A at the end I asked him how this research was funded and he said it was funded by Shell, but they weren't claiming any Intellectual Property Rights. He was also careful to state this was all "desk-based research." Another proposal discussed was the electrolysis of seawater. No proposals were discussed in any depth and, without exception, they seemed to be impractical due to the energy expended in operating such schemes. Some schemes aimed to "turn the clock back" in terms of claimed CO_2 levels. This lecture introduced me to the concept of "CO_2 removal cost" and one of the "magic figures" seems to be 40 dollars per ton. So, we get an idea of how people are thinking in terms of CO_2 and economics – all the terms and figures have been decided so that the issue can be "packaged" into discussions neatly.

This lecture also discussed "compensation issues" – how those adversely affected by any CDR (carbon dioxide removal) schemes should be compensated – directly, or following some kind of assessment/appeal process. It was said that CO_2 is a fungible commodity (essentially one which affects everyone). So, this led on to the concept of "carbon leakage." An example was given of the Drax Power station (largest coal-fired station in the UK) which simply could not make money if it had to pay the suggested CO_2 removal cost (tax). Again, all built on the unproven premise that CO_2 is the main problem.

Interaction Between Geoengineering and Emissions Mitigation
Dr Naomi (Nem) Vaughan of the UEA

She presented her research which was based on simple modelling of CO_2 levels which would result if CDR mechanisms along with SRM mechanisms (Solar Radiation Management) were to be employed. Again, her model assumed the global effects of the carbon cycle were well understood. Her model made projections many years into the future (up to the year 3000 or so!) No feeling was given as to what level of confidence these projections had – it seemed to be a case of "let's get a computer to produce all these nice graphs based on some figures we think might be right" and let's just talk about that for a bit.

In the Q & A following this session, David Keith (interestingly) introduced the notion of how things like attempts to do Solar Radiation Management could cause wars. He introduced the basic concept of "turning the knob" – geoengineering could end up being something that people could go to war over.

An engineer then questioned the lack of usage of error bars on the graphs – and how none of the graphs shown had them. He made a general criticism of how difficult it is to find discussion of uncertainties throughout the entire field of geoengineering and he politely requested that researchers make a greater effort to describe these. One of the other speakers acknowledged this, to some extent, but then argued that some uncertainties had been discussed in some papers.

Geoengineering: preparations and options for governance

Dr Margaret Leinen, Climate Response Fund, USA

The speaker stated she had worked in government and had also been involved with projects funded by venture capitalists. This presentation brought up points such as public involvement and consent – i.e. how can they consent to something they don't understand. There were no photos or graphs in this presentation. She mentioned the next big summit is Rio 2012 – where broader principles of governance would be discussed. She seemed to be worried about research being limited or stopped.

Public perception of geoengineering – knowledge, risk and acceptability

Professor Nicholas Pidgeon, Cardiff University, UK

This presentation was about public perceptions – it steered clear of the notion of "marketing" geoengineering to the public, but the sorts of information it covered could be used by those *wanting* to market it to the public.

This presentation was more engaging, though still troubling. The presentation said the geoengineering issue is polarised with "hope and fear" rather than "hype and hope." In the survey that this presentation was centred around, it was found that 50% of people had not heard of geoengineering and the more people learned about it, the less happy they were. Not surprisingly, it found that "natural processes" for geoengineering – afforestation, particular types of crop growth (to increase land albedo/reflectiveness) were favoured.

Geo-governance: assuring the future

Jonathon Porritt, Forum for the Future, UK

It should be noted that Porritt, like UK wildlife documentarian Sir David Attenborough, is affiliated with the group "Population Matters"[74] (formerly the Optimum Population Trust). Porritt's presentation was very light-hearted and poked fun at the whole issue. He did not make it clear whether he was advocating the use of geoengineering *and* he mentioned "Climategate" (which is covered in Chapter Three), though he sadly did not go into any detail. He said he trusted scientists more than he trusted politicians – as the latter are only concerned with short term thinking (to get re-elected). My impression was that he was stoking the fires of pointless debate. Porritt stated he was *not* in favour of a moratorium on geoengineering. He also used the phrase "a wholly new world order" where geoengineering is considered. It was rather an odd turn of phrase. He pointed out how businesses "play the game" very well. Porritt also noted the title of the conference and referred to the definition of "stewardship" (of the earth) which apparently comes from the words "sty-ward" – as in someone who looks after the pig sty. Again, there was no real discussion of scientific evidence – it just a light-hearted commentary.

Political dimension and perspectives

Professor Robert Watson, Department for Environment, Food & Rural Affairs, UK

He spoke of the vested interests in the issue, as well as government interests. He noted that the private sector in the USA "leads" government and the reverse is generally true in the UK. (But I would argue he simply has no idea of the extent to which he is right about the USA, and he does not really understand that this does happen in the UK too). He expressed a "pro-nuclear power" stance – with only a "possible role" for geoengineering!

Final Question and Answers Session

This session lasted about one hour, but all the questions assumed that CO_2-based climate change was real. Sir Crispin Tickell[75], a leading environmentalist said that there should be a "world environment organisation", though no details of what he was proposing were discussed by him or panel members. (This kind of idea is discussed more in Chapter Fourteen.)

I waited for about 30 minutes to speak and finally got my turn, just before the end (I have included a link so you can hear my comment at about 43:26 into the discussion segment[76]). I stated that I ran the website http://www.checktheevidence.com/ and that I was apparently in a "minority of one" in the room. I said I had watched the AGW issue being propagandised for 20 years and I was relieved that some people were speaking out. I also said I knew geoengineering was already in use. I then read out a segment from Eisenhower's 1961 speech, included earlier in this book (they tried to interrupt me). I then just said that if people were interested in what I'd found out, then they could come and see me on the pavement outside. I then left the room and got my leaflets[77] and other info ready. I waited for about 1 hour, as people left the building and I had a few conversations with people and gave out about 20 leaflets or so – to those who were interested. I also gave out a couple of DVDs. Some people were interested and stopped to talk for a couple of minutes, but it was cold and dark by the time I came out. I followed up with a few emails, but no one properly engaged with the evidence – preferring simply to ignore it.

Royal Society Meeting in 2013 – Next Steps in Climate Science

Figure 12 - Meeting in 2013 [78], which I attended.

In the 2013 meeting, I didn't even bother to make any comments, as there seemed to be no chance of breaking through the now fully reinforced "propaganda wall." Piers Corbyn of Weather Action[79] repeatedly challenged the speakers though, in a concise and civil manner. I spent some time with him immediately after the event and then communicated with him about the evidence in Chapter Thirteen of this book. He ignored this evidence.

Perhaps the title of the 2013 conference ("Next steps in climate science") is an admission that they will continue to ignore evidence, come what may, and help to further an agenda that we will be talking more about, also later in this book.

It was again the case that academics discussed limited sets of evidence and flawed models and convinced themselves what they think is correct/valid. In breaks between talks at the 2013 event, I approached two speakers and offered them a summary of the information which is near the end of this book. One of them refused even to take a leaflet from me. The other appeared to listen to me, but he did not follow up with any contact.

Of particular concern at the 2013 meeting was the final presentation by a certain John Ashton. He was a co-founder of E3G[80], a group described as follows:

> E3G is an independent climate change think tank operating to accelerate the global transition to a low carbon economy. E3G are the independent experts on climate diplomacy and energy policy. Our senior leadership has a combined 75 years experience advising Government, business and NGOs and a wealth of insight into what climate change means for societies.

They list among their donors Shell (Oil), the presumptively-named UK Department of Energy and Climate Change, and quite a number of NGOs and Government bodies. They also state the Rockefeller Foundation has funded some of their work[81].

Ashton is described as[82]

> ... an independent speaker, commentator and adviser. His activities range widely over politics, economics, diplomacy, and culture but his particular focus is climate change.
>
> From 2006-12 John served as Special Representative for Climate Change for three successive UK Foreign Secretaries, spanning the current Coalition and the previous Labour Governments. The UK Foreign Office pioneered during this time a diplomacy-led approach to climate change that came to be widely admired.

In a short presentation at the end of the day, Ashton told the audience that any scientists should basically realise they were going to have to work hard to get funding and he strongly implied that the best way to do this would be to make sure they made their "pitches" scary. In other words, Ashton, in my opinion, was not concerned about the truth of any causes for any climate change - he was more concerned with staying on the gravy train. He also implied that he could offer advice in "getting the message across" to people in an appropriate way.

After the meeting, I wrote to him, and some research assistants (listed on E3G's) website, and asked them how I could convey the information I had (which is now in this book) to more people, with the limited budget and means that I have. I never received a reply.

Council on Foreign Relations
This is another "think tank" which rarely gets a mention in mainstream media. Indeed, few people will have heard that Hilary Clinton said of the CFR, when speaking at a meeting in Washington DC[83] :

> Thank you very much, Richard, and I am delighted to be here in these new headquarters. I have been often to, I guess, the mother ship in New York City, but it's good to have an outpost of the Council right here down the street from the State Department. We get a lot of advice from the Council, so this will mean **I won't have as far to go to be told what we should be doing and how we should think about the future.**

The CFR has also published a number of reports about geoengineering and other articles mentioning this term. For instance, there is an article from May 2014 titled "Next Steps in Arctic Governance[84]." This is the same terminology used in the 2013 title of the Royal Society's Geoengineering meeting... Also, geoengineering is mentioned in a 2008 report about the "David Rockefeller Studies Program"[85].

Conclusion
So, this chapter has shown you a typical example of how the academic community is operating and how it will stone-wall anyone who wants to show them a different set of evidence. In the next chapter, we will look at one such set of evidence. Later, we will look at evidence which could (and should) make their "castle of lies" crumble into dust.

CHAPTER FIVE
"DON'T TALK ABOUT CHANGES IN THE SOLAR SYSTEM!"

"That is the essence of science: ask an impertinent question, and you are on the way to the pertinent answer."

Jacob Bronowski (1908–74) *The Ascent of Man*, Ch. 4

Space for Notes Below

Climatology vs. Astronomy?

Here, we will consider another likely cause of warming or significant climate change (for anyone who feels the need to say this has really happened). This has more likely a considerable set of independent and diverse data to back it up. I would like to emphasize, however, that this is *not* the most important section of this book and I am quite aware that believers of the AGW theory go to great lengths to try and explain away the evidence presented here. I can't really see how this is possible though. AGW believers are primarily focused on one tiny aspect of the earth's atmosphere – CO_2 content (which is only 0.03% of the atmosphere). They will also agree that water vapour is a much more "powerful" greenhouse gas than CO_2 – but they don't want to ban watering your front lawn or washing your car ... (I'm being facetious ...)

In general, I will just use some images and brief quotes – with references – and leave the interested reader to look elsewhere for the details, if they so choose. I will repeat that my main goal in this work is not only to refute the idea of AGW but also to show you how egregiously wrong it is, and *why* the idea was "dreamt up" in the first place.

Solar Activity

In 2003 NASA, a huge promoter of the AGW scam, posted an article titled "NASA Study Finds Increasing Solar Trend That Can Change Climate."[86] In it, they write:

> Since the late 1970s, the amount of solar radiation the sun emits, during times of quiet sunspot activity, has increased by nearly .05 percent per decade, according to a NASA funded study.
>
> "This trend is important because, if sustained over many decades, it could cause significant climate change," said Richard Willson, a researcher affiliated with NASA's Goddard Institute for Space Studies and Columbia University's Earth Institute, New York. He is the lead author of the study recently published in Geophysical Research Letters.
>
> "Historical records of solar activity indicate that solar radiation has been increasing since the late 19th century. If a trend, comparable to the one found in this study, persisted throughout the 20th century, it would have provided a significant component of the global warming the Intergovernmental Panel on Climate Change reports to have occurred over the past 100 years," he said.

Figure 13 - The Sun - with Sunspots!

Of course, this is somewhat heretical because the AGW theory needs to involve human activity – and no one can say the sun is influenced or controlled by human activity, so it won't surprise you to read a later quote from the same article:

> *Although the inferred increase of solar irradiance in 24 years, about 0.1 percent, is not enough to cause notable climate change, the trend would be important if maintained for a century or more. Satellite observations of total solar irradiance have obtained a long enough record (over 24 years) to begin looking for this effect.*

If it wasn't so sad it would be funny – NASA and other similar establishments give us photos, video and other data about CME's – Coronal Mass Ejections – enormous clouds of charged particles[87] – but little or no public discussion of the effect on our climate and weather of these highly energetic clouds – which regularly impinge on our atmosphere. Everyone knows, however, that thunderstorms involve *a lot* of electric charge[88] (the famous quote from the first "Back to the Future" film of 1.21 gigawatts is accurate). The same is true of other storm systems. Yet, this connection is never openly discussed. If you read to the end of this book, you might stumble across the most likely reason for this lack of public discussion.

The Maunder Minimum
Britannica.com defines the Maunder Minimum thus[89] :

> **Maunder minimum**, *unexplained period of drastically reduced sunspot activity that occurred between 1645 and 1715.*

This is important because of its possible or even likely link to the significant global cooling that occurred then. AGW promoters far too often ignore or want to explain away the enormity of the sun's influence. For example, a 1976 paper titled "The Maunder Minimum" by John Eddy[90], concludes there probably was a link between sunspot activity and "the little ice age."

> *The coincidence of Maunder's "prolonged solar minimum" with the coldest excursion of the "Little Ice Age" has been noted by many who have looked at the possible relations between the sun and terrestrial climate (73). A lasting tree-ring anomaly which spans the same period has been cited as evidence of a concurrent drought in the American Southwest (68, 74).*

The paper goes into quite some detail. Just consider the energy output of the sun (even though it is 93 million miles away) in comparison to the energy output of mankind, through industrial processes. Maybe it's just me.

The Maunder Minimum is relevant to research done by Jasper Kirkby in relation to climate and Solar Activity. Kirkby observed an apparent link between Solar Activity and the amount of cosmic ray radiation reaching the earth's surface. Kirkby was involved in work at CERN with regard to the effects of cosmic rays on cloud formation[91]. In 1998 he stated:

> *that the sun and cosmic rays "will probably be able to account for somewhere between a half and the whole of the increase in the Earth's temperature that we have seen in the last century."*
> [92]

The Medieval Warming Period (MWP)
This is generally disliked by the anthropogenic climate change catastrophists – who have a lot of evidence to argue with – when many diverse records show that the period between about 1000 and 1300 was as warm in many parts of the world as it is now[93]. Clearly, this warming cannot be blamed on industrialization! Lord Christopher Monckton has

written about this in some detail, as well as the contested "Hockey Stick" graph[94]. (The "Hockey Stick" graph is alleged to show how global temperature have risen sharply since industrialisation.)

It is interesting to note that an article about MWP on Britannica.com[95] provides a good overview of the evidence that the MWP was *real*, yet they write:

> The notion of a medieval warm period is highly controversial. Many paleoclimatologists claim that well-documented evidence for the phenomenon appears across the North Atlantic region, while others maintain that the phenomenon was global, occurring all over the world.

I contend that the main reason for the MWP being "controversial" is the intended result of the enormous amount of propaganda that has been promulgated since about 1990. The MWP is not controversial – it is inconvenient to the reigning dogma.

The reality of the MWP appears to show two important things – (a) global warming can occur as a result of something other than pollution and industrialised human activity, and (b) a warmer climate won't necessarily mean "the end of the world." Simply speaking, if the climate is warmer, then more water may evaporate from the seas and lakes, which should create more cloud which in turn would reflect more sunlight – and hence could start to reduce the temperature again …

Planetary Changes

If an accurate view of the possible causes of climate change is to be obtained, it must be recognised that ALL planets (with an atmosphere) in the solar system are undergoing climate change. Basic, referenced data is presented in this section and, once again, this should be *carefully reviewed* and *not ignored*. It can be seen that since the 1970s, significant *changes* (i.e. not specifically "warming" alone) have been observed and recorded on ALL planets of the solar system. The significance of this should be obvious. An article by Wm. Robert Johnston provides some additional comparisons to the ones shown below[96].

Venus – Changes in Composition of Atmosphere

An article posted on 23 Feb 1999, hosted on ScienceDaily.com states[97]:

> *"Our model shows Venus has changed dynamically in the recent past,"* said Bullock. *"Since Venus and Earth have a number of similarities, there are implications here for our own future."* An article by Bullock and Grinspoon regarding global change on Venus appears in the March issue of Scientific American.

It also says:

> *New computer models that indicate the climate of Venus has wavered radically in its relatively recent past may prove valuable to scientists tracking Earth's changing climate, according to two University of Colorado at Boulder researchers.*

Of course, the article does not mention the lack of human activity on Venus …

Figure 14 - Venus - composite Mariner 10 image

A quote from a 1995 article titled "Space Science – Solar System Exploration"[98] as part of the "Aeronautics and Space Report of the President states:

> *Venus has less sulfur dioxide, implying less volcanic activity, than in the 1970s.*

Another article "Night-time on Venus" dated 18 Jan 2001 reports[99]:

> *The green line intensity of Venus is comparable with that from aurora on Earth – a phenomenon produced as the terrestrial magnetic field interacts with the solar wind. But the atmospheres of the two planets are enormously different in composition, temperature and pressure. "A green line does not mean that a planet has an oxygen atmosphere because Venus has extremely low levels of oxygen", Slanger told PhysicsWeb. This result has implications for researchers studying the atmospheres of extra-solar planets.*
>
> *The team also hopes its findings may shed light on the apparent variability of the emission lines: the Russian Venera orbiters visited Venus in 1975 and found no sign of the green signal. "We do not understand how the variability can be this large", said Slanger, although the team speculates that the fluctuations could be connected with the solar cycle.*

Earth

We have already discussed the MWP and the Maunder Minimum. However, it appears that, for example, new forms of lightning such as Sprites, Jets and "mega-lightning" have been observed for the first time in the late 1980s and early 1990s. Consider this image from 24 July 2017, posted on SpaceWeather.com:

Figure 15 – The Gemini cloudcam atop Mauna Kea in Hawaii monitor storms approaching some of the world's largest telescopes. It often captures bright bolts of lightning lancing down to the ground below the towering dormant volcano.

It is also worth noting how much energy is in a single bolt of lightning – approximately a Gigajoule (1 billion joules).

SpaceWeather contributors have also seen changes in noctilucent cloud patterns over the years. Other high atmosphere phenomena such as the rare nacreous clouds have been photographed more frequently. Could these be manifestations of energy changes in the earth's atmosphere, in a similar way to what we have seen in the atmosphere of other planets in the solar system?

Figure 16 – Noctilucent clouds. Photo by Kristian Pikner, 13 July 2016, 23:57:49 [100]

Mars

An article on the Hubble Space Telescope website titled "Hubble Monitors Weather on Neighboring Planets" states[101]:

> To the surprise of researchers, Hubble is showing that the Martian climate has changed considerably since the unmanned Viking spacecraft visited the Red Planet in the mid-1970s. The Hubble pictures indicate that the planet is cooler, clearer, and drier than a couple of decades ago. In striking contrast, Hubble's observations of Venus show that the atmosphere continues to recover from an intense bout of sulfuric "acid rain," triggered by the suspected eruption of a volcano in the late 1970s.

In another article on the same site titled "Seasonal Changes in Mars' North Polar Ice Cap," we read:[102]

> "Particularly evident is the marked hexagonal shape of the polar cap at this season, noted previously by HST in 1995 and Mariner 9 in 1972; this may be due to topography, which isn't well known, or to wave structure in the circulation. This map was assembled from WFPC2 images obtained between Dec. 30, 1996 and Jan. 4, 1997."

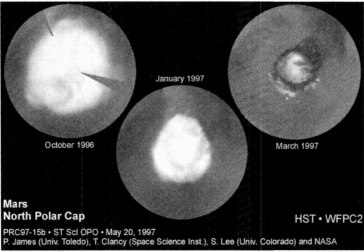

Figure 17 – Mars – 1990s image sequence from Hubble Space Telescope (HST)

Another quote from the 1995 "Space Science – Solar System Exploration" article[98] mentioned above, reads:

> The atmosphere of Mars is much clearer and colder than 20 years ago.

Jupiter

An article from *UC Berkeley News*, dated 21 April 2004 reports "Researcher predicts global climate change on Jupiter as giant planet's spots disappear."

> "According to Marcus, the imminent changes signal the end of Jupiter's current 70-year climate cycle. His surprising predictions are published in the April 22 issue of the journal Nature."[103]

Here, we can see very significant changes over a short period of time. The article continues:

> Marcus approaches the study of planetary atmospheres from the untraditional viewpoint of a fluid dynamicist. "I'm basing my predictions on the relatively simple laws of vortex dynamics instead of using voluminous amounts of data or complex atmospheric models," says Marcus.

Not surprisingly, it then adds:

> Marcus says the lesson of Jupiter's climate could be that small disturbances can cause global changes. However, he cautions against applying the same model to Earth's climate, which is influenced by many different factors, both natural and manmade.

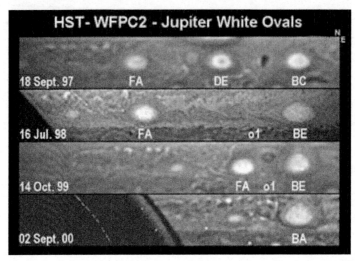

Figure 18 – Images from the Hubble Space Telescope show that between 1997 and 2000, two of Jupiter's three White Ovals disappeared. UC Berkeley's Philip Marcus predicts that the disappearance of these storm systems will be followed by a global climate shift on Jupiter within the next decade. (Images courtesy NASA)

An article was posted on NASA's science website on 3 March 2006 – titled "Jupiter's New Red Spot."[104]

> Backyard astronomers, grab your telescopes. Jupiter is growing a new red spot. Christopher Go of the Philippines photographed it on February 27th using an 11-inch telescope and a CCD camera:

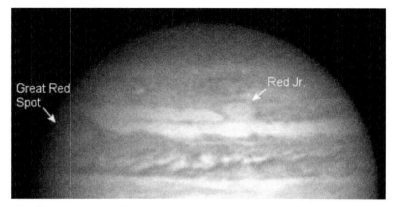

Figure 19 – A second Red Spot appears in Jupiter's atmosphere in March 2006.

In a 2008 posting on the Hubble Space Telescope Site, titled "New Red Spot Appears on Jupiter", this observation was confirmed – along with the appearance of a third spot.[105]

Figure 20 – Another Red Spot appeared in Jupiter's atmosphere in early 2008.

> *In what's beginning to look like a case of planetary measles, a third red spot has appeared alongside its cousins – the Great Red Spot and Red Spot Jr. – in the turbulent Jovian atmosphere.*
>
> *This third red spot, which is a fraction of the size of the two other features, lies to the west of the Great Red Spot in the same latitude band of clouds.*

It should be noted that the Great Red Spot on Jupiter has a bigger diameter than the earth. These are enormous changes.

An article was posted on NASA's science website on 20 May 2010 titled "Big Mystery: Jupiter Loses a Stripe"[106]

Figure 21 – 20 May 2010 – "Jupiter Loses a Stripe"

> *In a development that has transformed the appearance of the solar system's largest planet, one of Jupiter's two main cloud belts has completely disappeared. "This is a big event," says planetary scientist Glenn Orton of NASA's Jet Propulsion Lab. "We're monitoring the situation closely and do not yet fully understand what's going on."*

Saturn

A Wellesley College article report, in 2003[107] – "Saturn's rotation puts astronomers in a spin – Saturn's Equatorial Winds Decreasing, Spanish-American Team's Findings Raise Questions About Planet's Atmosphere."

> *The most commonly cited figure for Saturn's rotation period – 10 hours, 39 minutes and 22.4 seconds – was derived in 1980 from Voyager observations of radio waves generated by solar radiation hitting the planet's atmosphere. Yet Cassini has returned a result almost 8 minutes longer, a difference that defies easy explanation.*[108]

According to *New Scientist*, this article was re-published in *Nature* (vol 441, p 62)[109]. Another article has talked about an explanation for Saturn's Polar Hexagon.

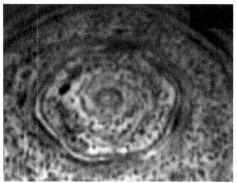

Figure 22 – an early photo of Saturn's hexagonal feature at the North Pole. (The Cassini probe has since photographed this in high resolution.)

What if there are "resonance wave" effects in the atmosphere creating centres of warming or cooling?[110] This could be what is happening on Saturn.[111] Few people talk about this and whether we might have equivalent effects, on a smaller scale, in our atmosphere.

Also, a huge storm was observed on Saturn – 29 December 2010[112]. What could be causing such enormous changes?

Figure 23 – Amateur astronomer's photo of the storm in December 2010.

This photo was posted on Spaceweather.com in Dec 2010 [113]

Figure 24 – Cassini Image of the storm in December 2010.

Uranus

A 1999 article from NASA's "Science @ NASA" website reports "Huge storms hit the planet Uranus."[114]

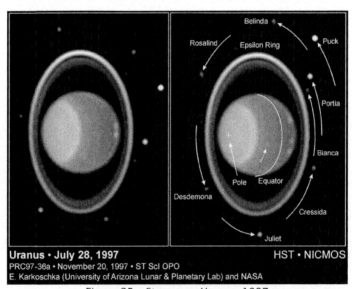

Figure 25 – Storms on Uranus, 1997

It then notes, "Infrared images from the Hubble Space Telescope reveal dramatic storm clouds moving in excess of 500 km/hr", and goes on to say:

> Early visual observers reported Jupiter-like cloud belts on the planet, but when NASA's Voyager 2 flew by in 1986, Uranus appeared as featureless as a cue ball. In the past 13 years, the planet has moved far enough along its orbit for the sun to shine at mid-latitudes in the Northern Hemisphere. By the year 2007, the sun will be shining directly over Uranus' equator.[115]
>
> Karkoschka, Hummel and other investigators used Hubble from 1994 through 1998 to take images of Uranus in both visible and near-infrared light.

A bright cloud was observed in 2005[116] (the referenced article was posted in 2011, however).

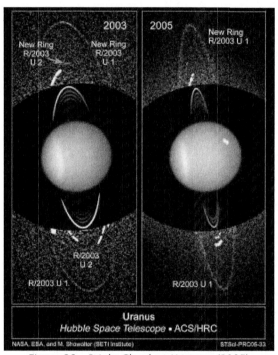

Figure 26 – Bright Cloud on Neptune (2005)

Neptune

Here, we have more Hubble photos showing significant planetary changes. As you will see below, the figures seem a little "off" to me.

> Seasons on Neptune occur for the same reasons as on Earth. The seasonal changes on both planets occur because their axes tilt slightly. Earth is inclined 23.5 degrees. Neptune is tipped at an even greater angle: 29 degrees. As both planets travel around the Sun, their southern and northern hemispheres are alternately tipped toward or away from the Sun.[117]

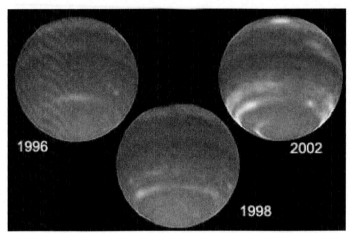

Figure 27 – Bright clouds appear on Neptune in late 1990s/early 2000s

But is this really correct? The time period covered is 6 years. However, Neptune's orbital period is 165 years – one quarter of this would be approximately 41 years ...

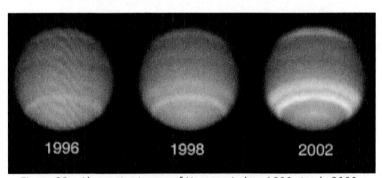

Figure 28 – Alternative image of Neptune in late 1990s/early 2000s

> Caption: A time series of images of the planet Neptune taken by the Hubble Space Telescope illustrate increasing cloudiness that is a hallmark of seasonal change. The growing bands of

> clouds in the southern hemisphere of the planet suggest seasonal change. Because the planet takes about 165 years to orbit the sun, the seasons on Neptune last more than 40 years.[118]

So, a season lasts six years – or 40 years? A further quote from the 1995 "Space Science – Solar System Exploration" article reads:

> HST images and spectra also revealed unexpected weather changes on Neptune ... There are unexpectedly rapid cloud changes on Neptune

Pluto

In an article from 2002, from MIT news[119], we can read "Pluto is undergoing global warming, researchers find."

> Pluto is undergoing global warming, as evidenced by a three-fold increase in the planet's atmospheric pressure during the past 14 years, a team of astronomers from Massachusetts Institute of Technology (MIT), Williams College, the University of Hawaii, Lowell Observatory and Cornell University announced in a press conference today at the annual meeting of the American Astronomical Society's (AAS) Division for Planetary Sciences in Birmingham, AL.
>
> The team, led by James Elliot, professor of planetary astronomy at MIT and director of MIT's Wallace Observatory, made this finding by watching the dimming of a star when Pluto passed in front of it Aug. 20. The team carried out observations using eight telescopes at Mauna Kea Observatory, Haleakala, Lick Observatory, Lowell Observatory and Palomar Observatory. Data were successfully recorded at all sites.

An article on Space.com, dated 09 October 2002 notes[120]:

> Though Pluto was closest to the Sun in 1989, a warming trend 13 years later does not surprise David Tholen, a University of Hawaii astronomer involved in the discovery.
>
> "It takes time for materials to warm up and cool off, which is why the hottest part of the day on Earth is usually around 2 or

> *3 p.m. rather than local noon,"* Tholen said. *"This warming trend on Pluto could easily last for another 13 years."*

Well, I guess that explains it! On earth, it takes a few hours to get to the hottest temperature, and this happens after the sun is at its most intense and it just takes 13 years on Pluto (a much smaller world and much colder place, so shouldn't the "thermal inertia" be smaller ...?)

Local Interstellar Medium

Few people have heard of the Interstellar Medium. Britannica.com defines it as follows:[121]

> **Interstellar medium**, *region between the stars that contains vast, diffuse clouds of gases and minute solid particles. Such tenuous matter in the interstellar medium of the Milky Way system, in which the Earth is located, accounts for about 5 percent of the Galaxy's total mass.*

Some researchers have tried to investigate it and also proposed that changes in the density or composition of this medium might affect the earth's climate[122].

> *Research on the properties of the Local Interstellar Medium have been carried out in scattered periods beginning in 1978. The NASA Space Physics Division has shown a persistent pernicious bias against work on the effects of the neutral gas in the LISM in the United States, from the time of the formation of the Division. The dominant role of neutral hydrogen in the formation of the termination shock in the collision of the solar wind with the LISM has only recently been recognized by the particles and fields research community, which has been supported primarily by the Space Science Division. The most important contributions to research in this program are papers (48), which presents a calibration independent method of determining absolute LISM density, and (89), which presents the first evidence for a large increase in the LISM neutral atomic hydrogen density from Voyager measurements of the 50 AU region, suggesting the approach to the termination shock (89). See 19, 20, 21, 48, 64, 82, 89.*

This is never brought up as an issue – which is another example of how AGW theory promoters cherry-pick evidence to ram down everyone else's throats and ignore almost all other pertinent evidence.

Summary

AGW believers and scammers will, of course, want to just say we know far more about the earth's atmosphere, due to a wider range of measurements. However, they cannot now deny observed changes on *all* other planets in the solar system – which are millions of miles away, yet still those same changes are significant enough to be clearly observable from earth.

A Possible Mechanism Behind These Changes?

Some people such as Richard Hoagland and David Wilcock (who have their own websites with, in my view, a mixture of good and bad material) have proposed that a "different physics" is behind these planetary changes. Perhaps this physics could explain the following facts (check them):

- Olympus Mons, 27 km high volcano on Mars – latitude 19 degrees
- Solar Maximum – most sunspots occur at latitudes of 19.5 degrees
- Red Spot on Jupiter 19.5 degrees.
- Big Island of Hawaii – latitude 19 degrees
- Dark spot on Neptune – latitude 19 degrees
- Alpha & Beta Regio – Venusian volcanoes – latitude – 19.5 degrees.
- Strongest El Nino currents occur on latitude – 19 degrees.

What if some "wider scale" energy system is causing all these changes (including the sun's unusual activity in the last 10 years[123])? If the same energy system is impinging on the earth, how would changes in it affect our own climate?

An Electrically Charged Solar System/Universe?

Conventional science contends the solar system is electrically neutral. In the "Thunderbolts Project", David Talbot and Wallace Thornhill also provide a large body of evidence that there is an overall "shaping of the universe" by the electric fields associated with matter, and not the force of gravity[124]. They, too, do not believe in AGW theories[125]. They have explained many astronomical phenomena related to galaxy formation, comet data, solar events and so on as being the result of charge movements between stars, planets and other astronomical bodies. Their research is highly recommended. (But they, too, will also not properly discuss evidence I will cover later in this book.)

CHAPTER SIX
PERSISTENT JET TRAILS/CHEMTRAILS

"We must examine our own 'difficulties with the truth' and we shall discover that we too have cause for regret and shame."

Christa Wolf, German writer.

Space for Notes Below

SRM and Trails

In Chapter Four, we looked at some proposed schemes for Geoengineering. One class of schemes was Solar Radiation Management (SRM) – a way of affecting the amount of radiation reaching the earth's surface. Reducing the amount of radiation reaching the surface is perhaps one way, according to geoengineering proposals, to reduce the temperature of the planet.

Some people insist that such geoengineering schemes are *already* in operation. They state that we know this because of the large numbers of aircraft trails that can be observed in the sky on some days. We will now examine this evidence and see where it leads us.

Watch the Skies …

This section documents some the strange effects observable in relation to aircraft trails seen around the world. Mainstream media and academics generally don't study this phenomenon nearly as much as they should, and so anything postulated or claimed about any effects of the trails are (usually) too easily dismissed as "only being of concern to paranoid conspiracy theorists."

However, I now challenge the reader to look more deeply – and observe some of the troubling strangeness related to this phenomenon.

Many people are very much concerned with their lives "at ground level" and so it is quite a rare thing for them to look at the sky. This is not all that surprising when one considers that meteorologists, pilots and astronomers only make up a small proportion of the total population.

It is also not that surprising that people think little or nothing of the pollution of the air – it is an accepted, if unwelcome, part of modern life.

Is it any wonder then, that when people actually notice trails in the sky, they automatically think "oh yeah – those are just aircraft trails", and carry on doing whatever they are doing. However, when one actually stops, thinks and begins to study these trails more closely, a once clear "black and white" picture quickly fades from view.

If one studies the physics of the vapour trails of aircraft, the basics would seem to be fairly straightforward. In fact, the basics are something we often personally experience, at least in the United Kingdom, on every cold winter's day. On such days, when we breathe out, we can see our breath. It's one of those signs that "winter is really here." What causes our breath to become visible? Very simply, it is that our breath is warm and the winter air is cold. Tiny droplets of water vapour condense out of the warm air to form "clouds of visible breath", before the warm air quickly cools and the "clouds" disappear again.

It is a very similar process that is happening about 30,000 feet in the air, when hot exhaust gases from jet engines heat the air. Water droplets condense out of the cooler surrounding air and form a *contrail* – an abbreviation of *condensation trail*. I used to watch aircraft trails when I was a child and I remember seeing how the almost solid-looking lines of "stuff" would slowly fade into wispy curls, then disappear completely. Being as curious as I was, I probably observed this process through binoculars on more than one occasion. The trails would become invisible after perhaps one or two minutes.

On the Trail of the Trails

Looking at more recent aircraft trails, there seems to be a general trend that many of the trails no longer disappear in such a short time period. Indeed, in researching a little into these aircraft trails, I came across an observational study, which was done in 2002, by Amy Foy at Lancaster University (UK)[126]. Here, a classification of the type of Aircraft Trails observed was used:

1. "Persistent and Dispersed" (they hang around and spread out).

2. "Persistent and Non-Dispersed" (they hang around but don't spread out).

3. "Non-Persistent and Dispersed" (they don't hang around, but they do spread out).

4. "Non-Persistent and Non-Dispersed" (they don't hang around and they don't spread out).

The Lancaster study does not attempt to explain why some trails should be persistent or seen when dispersed, but it does show that someone else has observed these trails enough to see that some of them do persist for more than five minutes.

There is an article with similar categories of trails on one of NASA's websites[127]. This page has only three small images of contrails on it (reproduced below). Please compare these to images shown later in this section.

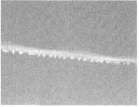

Figure 29 – NASA's contrail photos.

These are labelled "Short Lived", "Persistent (non-spreading)" and "Persistent Spreading"

Aircraft Fuel – Burn, Baby, Burn

Before we explore some of the chemistry of the burning of kerosene (aircraft fuel), let us stop and think for a moment. If, on a cold day, we breathed out, and our clouds of breath hung around for several tens of seconds or even minutes, would we regard this as unusual?

If aircraft trails are visible for several minutes, there must either be some component in them that is visible when cool or some visible compound must be forming in the atmosphere, following a chemical reaction of some kind. Alternatively, the conditions under which normal condensation of water vapour happens must change on some days. Let us explore these ideas.

Kerosene is classed as a "hydrocarbon" – it contains mainly alkanes – which are made up of carbon (approximately 85%) and hydrogen (approximately 12%). There are some other compounds in kerosene which contain nitrogen and sulphur (approximately one percent or two percent each respectively). When kerosene burns, therefore, it can only form compounds that contain elements that were originally in the kerosene, or in the air it burns in. Not surprisingly, then, the main compounds that form when kerosene burns are:

- carbon dioxide (the gas we all breathe out)
- Sulphur Dioxide (in small quantities – a toxic, greenhouse gas, which mixes with water to form acid rain – sulphurous and sulphuric acid)
- Carbon Monoxide – a toxic, flammable gas, responsible for some deaths which happen when gas heating equipment is faulty.
- Water.

When we look at each of these compounds in turn, we find that they are *all* colourless. So, when kerosene burns, it would seem that the only visible thing we should see in the sky is the condensation – which, like our breath, should disappear in a few tens of seconds. Indeed, when a jet takes off, we can see that only colourless compounds come out of back – all that we see is *"hot air."* There are no sooty or reflective compounds coming out as the jet races down the runway. Whilst these observations may not be true of all the jet engines that are currently flying, it should be true of all those used on regular flights, otherwise they are faulty.

So, whenever we see a contrail lasting for more than a few tens of seconds, we should, at the very least, be curious, and wonder what is causing this to happen? When we see a lot of these trails together, we should become very concerned. They should not be there in the first place, but accepting the fact they are, we should realise they are a very visible form of pollution, which few people seem to be paying attention to.

Tic-Tac-Toe or Noughts and Crosses Grid?

In about 2005, I started observing these trails more closely and found that, like many of those observed in the Lancaster University study, they persist for quite a few minutes. I made one or two time-lapse videos with a Webcam. I then witnessed an extraordinary sight on 10 June 2005 at 9:45 in the evening. A photograph I took of this is shown below. I sent it to a popular US talk show and they posted it on their website and then I received about 30 messages in response to it!

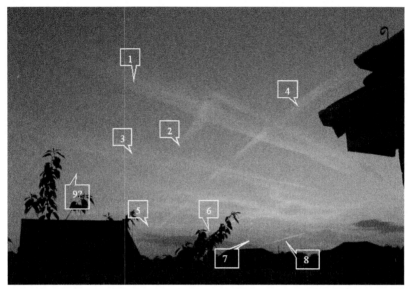

Figure 30 - Trail Grid photo (10 June 2005) and annotations by Andrew Johnson

A few of these e-mails thought that the pattern in the photograph was not that surprising – maybe just coincidence, but the majority of the messages were similar to those shown below:

> *"I live in Nebraska and this has become a familiar scene here. It boggles my mind how few people seem to take any notice."*
>
> *"In Los Angeles, we get them more than 80% of the year. Wish I knew what it was all about because I, like many others, think I know what's up but it's like taking a stab in the dark."*
>
> *"We get these over our house in Rochester WA several times a year. I've taken pictures that look just like yours."*
>
> *"Those are chem. trails and you need to understand what they are. Please contact me unless you have already been contacted by the 'educate yourself' website."*

Ordinary *contrails* from modern airliners should, as illustrated with the chemistry and physics outlined above, normally disappear in less than about two minutes, but many of them persist for many minutes or even hours. Depending on what you read, and who you believe, this is classed as "perfectly normal." All official sources state their formation is explainable as "contrails" – yet there is no consistent scientific explanation available which covers all the observed anomalies – such as grids, circles, two identical aircraft on the same day within one hour of each other with one leaving a long persistent trail and the other not (more on this later). Other data I have collected suggests that the reasons the trails persist, in some cases, cannot be explained by "the prevailing conditions" – which we will also cover later in this chapter. Please review the pictorial data and use your *own observations*.

Persistent Jet Trail/Chemtrail Phenomenon

On the internet, there is a whole "ecosystem" devoted to chemtrails and there is much misunderstanding and false information being circulated, which we will discuss later.

Since the mid-late 1990's, people around the world have observed what have become known, correctly or incorrectly, as "chemtrails." Mainstream science and commentary mostly considers these trails to be a normal result of everyday air traffic movements – i.e. they are purely and simply condensation trails formed as a result of burning kerosene. Others maintain they are part of a secret, clandestine "spraying programme." A researcher called Clifford Carnicom has proposed the following possible reasons:

- To help create environmental or climate changes (geoengineering)
- To introduce biological materials to affect humans or agriculture (toxic spraying?)
- For "military purposes"
- To change the electromagnetic properties of the atmosphere
- To cause geophysical or global effects
- To enable operation of exotic propulsion systems

At this point, the reality of the phenomenon is clear, but not the cause. Some claim the trails are being created through the use of fuel additives and some claim there are aircraft in operation that have a separate spraying system installed. Some people claim to have photographed additional nozzles on aircraft, but in most or all cases I am aware of, these have been shown to be for other purposes such as crop spraying or science research projects (there are some examples on the "metabunk" forum[128] such as a study of a story titled "Exclusive: Leaked Photos of Chemtrail Dispersal System"[129].) It is quite obvious that fuel additives would affect the performance of the engines/aircraft and would surely, over long periods of time, cause a residue to build up.

I have been watching the trails from the same location in Derbyshire (my house in a semi-rural area) for 12 years and have not personally felt any toxic effects, or seen any obvious decline in the flora and fauna around me.

What's in these supposed chemtrails, then ...?

Many people say barium and aluminium compounds are contained in the trails (we will revisit this later) and some even say that when the planes go over and leave heavy trails, their health soon deteriorates. Some tests of water and soil samples seem to show elevated levels. The evidence is patchy and unclear and in all cases I am aware of, no one has been able to prove that, if such compounds have been found in their water, that the source of them is from a plane which has flown over their location.

Clifford Carnicom has claimed to find various compounds in the trails. However, conclusively proving these compounds came from "chemtrails" is difficult. Carnicom has also reported finding Erythrocytes[130] (like red blood cells) – a biological component. However, it is not clear how he has confirmed the provenance of his sample – and proven that it came from a plane flown in the stratosphere.

A Website in the UK called "Chemtrails Project UK" has even been selling rainwater testing kits and collating and mapping the results[131]. However, these results seem to be quite inconsistent, and have not been shown to correlate with any aircraft movements, so are inconclusive.

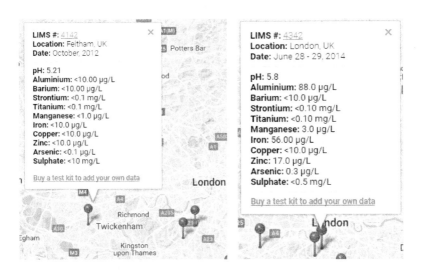

Figure 31 – Sample Water test results from "chemtrailsproject" website.

There are also claims by one or two scientists regarding elevated levels of aluminium or barium in soils affecting food production. One such claim is made by Francis Mangels (35 years Federal Scientist, USDA Soil Conservation Service US Forest Service[132]). Here is a partial transcript of a discussion with him.

> Ok, well when you start throwing these oxides, like aluminum hydroxide, barium oxide, boron, manganese oxide, these things, it raises the pH of the soil. Now, like I said, I've got the 1983 soil data for my area and I know which type of soil is naturally high in pH, and soils should be running somewhere between 4.5 to 6.0 and since the chemtrails have been started, pH has gone sky-high, about 10-20 times more alkaline.

He later says

> Well, the most dramatic thing that has happened in California is our tomato crops just went to pieces. We raised the pH so high in California that the bottom fell out of the tomato growers, and now they are raising other crops instead of tomatoes, that's one major thing."

The referenced article (above), however, shows that Mangels' comments about tomato crop yields aren't really true, according to data published by the State of California. It should also be borne in mind that aluminium compounds can be found in significant concentrations in some clays.[133]

A few years ago, I had some correspondence with Rosalind Peterson, a retired Crop Loss adjuster,[134] regarding the chemtrail issue and what might be being sprayed. In a 2006 News Broadcast[135] she stated:

> What I found was unusual spiking since the early 1990s in barium, aluminum [in the water supply of Northern California – says the news reporter].
> Q: What's the significance of the barium and the aluminum?
> A: I went and got all the tests – and all these things in the same test would be up – way over state and federal standards – these had to be airborne – because how could it get to such diverse regions of our county? ... We have jets going in every direction... X's, east, west, north, south – circles. There is something going on in the air – I don't know exactly what, but I think there's some experimentation ...

However, in 2012, she stated: [136]

> When it comes to proving what the jets are releasing, I don't have the documentation, and I don't have a single study, I don't have a single solitary verifiable evidence that the jets are releasing anything except military releases of aluminum coated fiberglass by military aircraft.

Jim Lee seems to agree – he has done some excellent work using high quality data, which he has published on his comprehensive and technically virtuosic website "climateviewer.com." There, he writes[137]:

> The entire "chemtrail conspiracy" boils down to one thing: intent. The problem with searching for intent is that it leaves people looking for a "smoking gun" or a "whistle blower" or a note from "the guy" who ordered planes to "spray" the world. This is actually a search for a "straw man" and likely designed to waste your time. If you want to end planes making clouds, all the evidence you need is in this article and all that is needed is proper protest at the FAA, EPA, and ICAO in my humble opinion. Are "rogue geoengineers" intentionally spraying material to create clouds that cover the sky or is this just a dirty, unregulated industry doing what fossil fuel industries always do, polluting? Based on the evidence we can show that the military has intentions to use carbon black dust to modify the weather for warfare purposes, and commercial aviation produces tons of carbon black dust.
>
> Due to the overwhelming amount of propaganda surrounding the topic of chemtrails, I have researched for more than three years to "clear the air" on chemtrails as a part of something I dubbed "Operation Clarity" and this is my final report. The information I present is based on information I have personally reviewed and is subject to change based on new evidence to the contrary.
>
> "There are three sides to every story. Your side, his side, and the truth."
>
> Conspiracy believers claim that chemtrails are a secret program that do X, Y, and Z. Debunkers say chemtrails are contrails and are completely normal (i.e. harmless). The truth is that artificial clouds are destructive to nature, harmful to

> health, and there is nothing "normal" about fire breathing metal tubes spewing nanoparticles at 30,000 feet. Despite more than 60 years of jet aircraft "accidentally" geoengineering, scientists are now wanting to legalize global weather control.

Again, Jim Lee does not discuss all of the evidence presented – nor the most important evidence of geoengineering that is included later in this book.

So, overall, I am now currently less convinced than I used to be about the validity of statements which declare the existence of a global, clandestine toxic spraying operation. Having said that, there have been disclosed instances of spraying the population with toxic substances. For example, between 1953 and 1964 top secret trials were carried out over East Anglia, UK[138] with a chemical concoction of zinc cadmium sulphide to simulate how a cloud would disperse biological agents. The unsuspecting population was sprayed covertly with the poisonous compound at least 76 times. A story in the UK *Guardian* from April 2002[139], discusses a

> "60-page report [which] reveals new information about more than 100 covert experiments. The report reveals that military personnel were briefed to tell any 'inquisitive inquirer' the trials were part of research projects into weather and air pollution."

In the USA, spraying of "biologically inert" gases into the air was disclosed in Oklahoma City 2003, in what could be called a "bio-terror simulation experiment"[140]. This experiment was conducted using a "combined budget from the U.S. Department of Energy, the U.S. Department of Homeland Security and the U.S. Department of Defense – Defense Threat Reduction Agency and other participating federal agencies" of $6.5M.

"Here Are Some Contrails ..."

Sites such as contrailscience.org [141] and metabunk.org [142] insist, perhaps partly because of photos like the two below, that there is no mystery to these trails. They seem to take pride in "debunking" the phenomenon – and several related ones. (Not surprisingly, these sites don't properly address the evidence laid out here and just casually ignore things either as coincidence or just as "nothing to see here – move along!") However, these sites are both measured and polite in their analysis and criticism and I have referred to these sites a number of times in my own research, and in compiling this book.

Figure 32 – *B-17 Bombers over Europe, 1943. "Lots of Ordinary Persisting Contrails, but no Chemtrails"* [143]. **But remember, regular civilian aircraft aren't meant to fly in formation in the way military missions or exercises are...**

In April 2017, someone gave me a copy of *Clouds and Weather Phenomena* by C. J. P. Cave[144] (Cambridge at The University Press, 1943). I don't think there were many jets flying around like this in 1943 – so it was probably from a prop-plane, but the reasons for contrail formation aren't all that different, it seems.

Figure 33 – Condensation trails from Fig 42 of the CJP Cave Book.

Persistent Jet Trails/Chemtrails

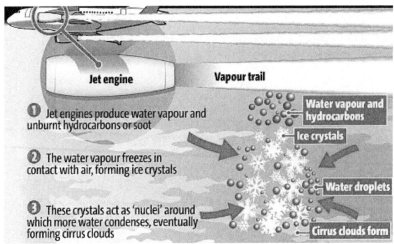

Figure 34 – Graphic about Contrail and cirrus cloud formation

Chemtrails are just contrails, right? Some people prefer "aerosol" or "persistent jet trail" (PJT) as chemtrails are often automatically dubbed a "conspiracy theory." They are real, not theoretical – and you can observe the anomalies for yourself.

As of the time of compiling this book, I have now been researching the PJT or chemtrail phenomenon for about 12 years. I am still waiting for some person to give me a sensible explanation for something artificial that I can see weekly or even daily. It seems there is very little direct information available outside personal observation and the odd rumour. I also know that disinformation is actively being promulgated, as I will show later in this chapter.

Reasons to Question the "Contrail" Explanation

Visibility of Trails on Satellite Photos

Jeff Challender (deceased) suggested that the mass of water vapour contained in a standard contrail would be tiny, and certainly not observable from 150 miles up in space and yet, we can see the trails on many satellite photographs, such as the one shown below. How are these trails persisting long enough to be photographed from space like this?

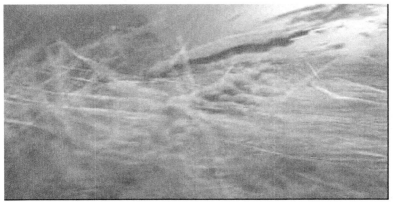
Figure 35 – Contrails seen from space.

This seems like an awful lot of persisting trails – making what amount to artificial clouds …?

How should we explain this pattern, for example from May 26, 2012?[145] (This one got into the *Daily Mail* – a bastion of media psychological operations?)

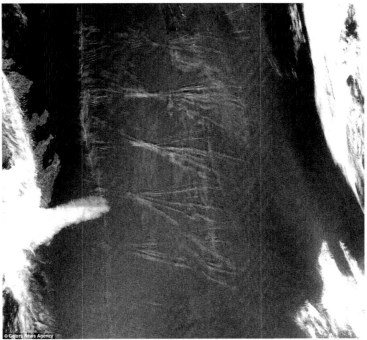
Figure 36 - Jet trails: NASA's satellite captured the jet trails, pictured, criss-crossing the Atlantic on May 26, 2012

I would suggest that it is debateable whether trails should be visible, to the extent they clearly are, on satellite photos. Of course, the resolution of cameras has increased over the years, so finer detail can be photographed, but the persistence of the trails is still, often, inexplicable. Similarly, any geometric formations, as shown in some of these photographs, are also harder to explain. Why do the trails form a kind of "fan" shape and connect with the trail which runs nearly vertically?

Also, as we shall see later, the frequency of appearance of trails does not bear any noticeable relationship to levels of civilian air traffic.

"Broken" Trails

In many cases, instances of "broken" trails are seen – and these "breaks" are also persistent. Although I am sure someone will tell me that this is to do with varying conditions in the atmosphere, there seem to be some occasions where this is difficult to accept. In some cases, the breaks in these trails seem to be deliberate – perhaps to form some kind of grid – the more that this is observed, the harder is appears to be to explain.

Figure 37 - Broken trail seen in Borrowash, UK

Case Study – Many trails seen around the UK, and from space – 18th October 2007

Over the years, I have observed that on certain days, there seems to be a large increase in the number of observed trails. In this section, I present a selection of photos that were all taken around the UK on 18th October 2007 [146]. Additionally, we see a large number of trails on a satellite photo.

There still seems to be no good explanation as to what atmospheric conditions cause this.

Figure 38 – Derby trails on 18 October 2007

Figure 39 – Hertfordshire trails on 18 October 2007

Figure 40 - Leeds - trails on 18 October 2007

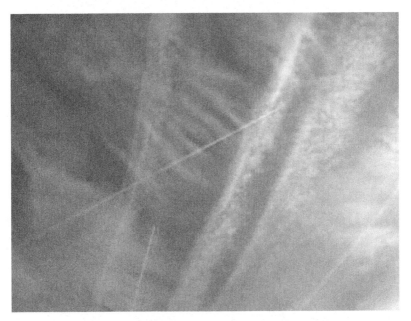

Figure 41 - Keighley trails on 18 October 2007

Shot of England from space on 18 October 2007

Europe_2_01 – Date: 2007/291 – 10/18, [147] True color – Satellite: Terra – Pixel size: 1km

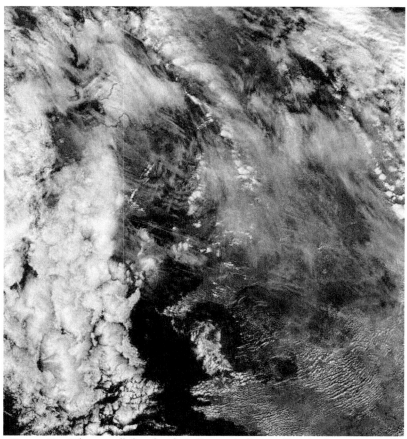

Figure 42 – Why are all the clouds in France in lines – like ripples on a pond?
(Photo now archived and not easily accessible [148])

Grids, Circles and other Trail Phenomena

The regular appearance of these is troubling and I have yet to read a convincing explanation.

In the photographs below, mainly from around the United Kingdom, a number of circles and grids of trails can be seen. There seems to be no good, clear explanation for this and, to my knowledge, military exercises have not been proven to be the cause of any of these "displays."

Figure 43 - 10 June 2005 - Borrowash, UK

Figure 44 - 8 August 2005, 13-04 Embsay, Yorkshire (Damian Sheeran)

Figure 45 - 21 January 2007, Humberside, UK

Figure 46 – 03 August 2007 – Borrowash, UK

Figure 47 –15 April 2008 – Pyrenees, France/Spain (Alan Cruickshank)

Figure 48 - 21 February 2009 - Borrowash, UK

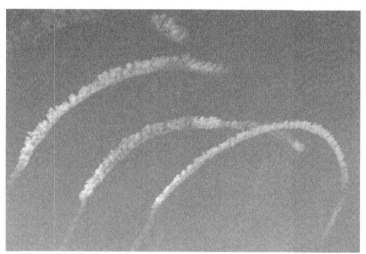
Figure 49 - 16 March 2010 - Blaneau Ffestiniog, Wales, UK (Dave Ellis)

Figure 50 - November 2010 - Edinburgh, UK

Figure 51 - 29 November 2012 - Hendon, UK

Figure 52 – 16 January 2012 – Kidderminster, UK

Figure 53 – 16 January 2012 – Louth, UK (M Kheng)

In the photo above, from the 16th of January 2012, it is alleged that a "NATO plane" created these circles of trails [149], though if this was true, the full purpose of the exercise that created them is not clear. An article in the *Louth Leader* claims "the aircraft was a NATO Sentry E3, a surveillance plane, which was on a sortie completing a standard UK orbit."

Figure 54 - 15 March 2013 - Grimsby, UK

Figure 55 - 04 August 2013 - Ashtead, UK

Figure 56 – 18 August 2013 – Ashtead, UK

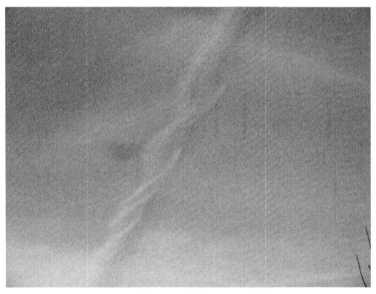

Figure 57 – 10 January 2014 – Lancashire, UK

Figure 58 – 19 October 2014 – France, Photo by Chris Marsh

PJT/Chemtrail Disinformation

In researching the chemtrail topic it has become clear that disinformation has been actively circulated, so I will briefly discuss a couple of examples here.

An article originally published by Ted Twietmeyer on 11 May 2008 shows the image and caption below: [150]

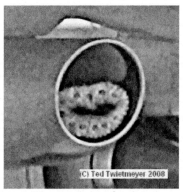

Figure 59 – See below for description

> *"Fig. 5 Spray pod exit port, brightness enhanced since it was in the shadow of the sun. The brush-like object would provide the maximum surface area to create an aerosol. Note the dark baffle which will restrict airflow completely through the aerosol device."*

He later had to correct his mistake – The "pod" in question is likely either a refuelling pod or a smoke generator[151].

Figure 60 – Refueling Pod

Another piece of disinformation that seems to have been widely circulated, was a supposedly "leaked" photo showing tanks of fluid "inside a chemtrail plane"!

Figure 61 – Inside a chemtrail plane!! Yes really! (Disinformation)

What you are actually looking at is an aircraft under test. Tanks simulate airframe load (fluid can be pumped around to test this).[152]

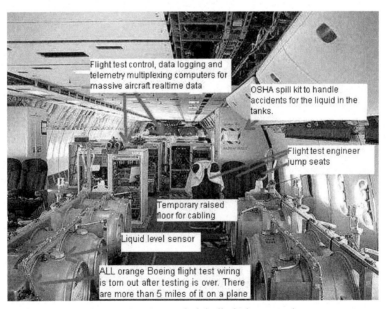

Figure 62 – Correctly labelled photograph

Summary

In this chapter, then, we have seen photos of peculiar trail phenomena which don't seem to have a clear or convincing explanation in all cases. Additionally, we have seen other images which "chemtrail" bloggers and activists have wrongly promoted as being proof of some type of aerial spraying being in use (and more examples can be found online, without too much difficulty). Hence, it appears there is an unusual, real phenomenon to be witnessed which is important – because someone is deliberately promoting false information about it. We will cover further evidence of the importance of the PJT phenomenon in the next chapter.

CHAPTER SEVEN
STUDYING CONTRAILS/PJT

"Who has seen the wind? Neither you nor I:
But when the trees bow down their heads,
The wind is passing by."

Christina Rossetti (1830–74)

Space for Notes Below

Asking the Questions ...

In 2007, I sent a 20-page report I had compiled about the PJT/chemtrail issue to several UK bodies – including the Department of Environment, Civil Aviation Authority, Royal Air Force, Greenpeace and the Worldwide Fund for Nature – and others. I also sent a copy to the UK Department for Transport. The responses I got did not explain the observations I've described above. However, one useful response came from a certain Roger Worth of the DfT and we corresponded briefly. He sent me a link to a copy of a paper called *Formation, Properties and Climatic Effects Of Contrails* (Schumann, 2005) [153] by Professor Ulrich Schumann of the Institute of Atmospheric Physics. At that time, the paper was only two years old – and now it is 12 years old ... So, please understand that this section refers to this paper only.

I studied the 2005 Schumann paper in some detail and could not really find anything that was specifically relevant to the data in my own report – such as how grids and parallel lines are formed. The Schumann paper talks about contrail formation being possibly linked to cirrus cloud formation, but states there is no proven link between them. It does indeed discuss persistent "contrails" but does not explain why they form. The duration of their persistence is not discussed in detail, or with any empirical data. In particular, my attention was drawn to two figures in the report: the standard contrail duration of maximum two minutes (with which I have no argument) and also the discussion of regions of ice supersaturation. It states that ice supersaturation in the atmosphere may be the cause of persistent contrail formation, but no firm link is documented or established. Indeed, a figure of 150 km is quoted (Section 4, just under Figure 5) for the maximum size of a region of ice supersaturation.

This still doesn't explain why trails persist. Also, I measured a trail over 300 km long – which presumably would fall into the "unexplainable" category based on figures given in the Schumann paper.

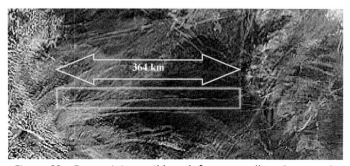

Figure 63 – Determining trail length from a satellite photograph

Figure 63 above begs a simple question. How is it possible for trails to persist for so long that they form long lines? Look at the trail marked in a separate photo.

This trail is 172 pixels long – this means that at 2 km per pixel, the trail is about 364 km long. (A small adjustment may need to be made due to the distances above ground. If the ground resolution is 2 km/pixel then at a height of 30,000 feet, the resolution would be maybe 1.9 km per pixel). If we assume it was made by an aircraft similar to a 757 or an Airbus A320, and we assume the plane was travelling at 500 mph for the time the trail was forming, this means that the trail persisted for at least:

$$364 / (500 \times 8/5) = 0.455 \text{ hours} = \underline{\textbf{27 minutes!}}$$

(and it could be longer, since the satellite photograph may have been taken *after* the trail had formed.)

Some discussion of LIDAR measurements is included in this report, and this is quite interesting, but inconclusive. There are no ordinary (optical) photographs in this study and there are no time-lapse studies.

For example, in the Schumann paper – Section 6, Figure 6, we have: *Compare with satellite photo from Feb 4th, 2007.*

Figure 64 – Shumann paper – one of only two images in the paper. Compare to other satellite photos of trails.

Elsewhere, we can find other satellite photos showing many PJT's – much more clearly – for example, this one (or refer back to ones shown in the previous chapter).

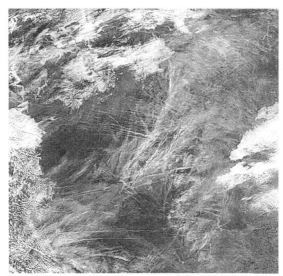

Figure 65 – One that is more representative of the problem – Rapidfire Website 4th February 2007 [154]

As I will show you, these studies can be made with cheap and simple equipment and are useful for gathering quantitative raw data. Coupled with other methods for gaining information about the state of the upper atmosphere, this could form the basis of more useful study. Of course, as I am a private individual without access to research grants and resources, I have progressed this as far as I can.

Throughout the Schumann paper, statements are made which include the word "may" – indicating speculation – and often times, the statements aren't supported directly (in the paper itself) by observational evidence. For example, Section 7, paragraph 5 reads

> *The ice formation processes are very complex and not yet finally understood [5, 6, 92, 98, 99]. The changes in concentrations of ice nuclei (such as aircraft soot) may cause an increased cirrus cover but may also cause a reduced cirrus cover, so even the sign of this effect is presently uncertain [100].*

Similarly, in Section 5, paragraph 2

> *At present, only a few exploratory studies have dealt with the later stage of the persistent contrail dynamics which depends*

> on the mesoscale atmospheric flows with rising or sinking motions of turbulent or wavy character and on shear, radiation and ice particle sedimentation. A vertical shear in the wind perpendicular to the contrail causes a contrail spread **which may reach several kilometres within hours** [73-76].

At the start of Section 9 we read:

> "The climatic impact of contrail cirrus is not known."

On a global scale and/or long term this may be true, but I have documented the effect, as have others, on short term, localised climate change – where a haze develops and sunlight levels drop. This is a known, effect which is repeatedly observed and I have even recorded at least one time-lapse video of it happening.

In conclusion, the paper lists an impressive number of references, but sadly it completely fails to explain the type of effects seen with trails that have been documented by hundreds or thousands of people across the world.

"First Published Study on 'Chemtrails' Finds No Evidence of a Cover-Up"

In November 2016, a study was published titled "Quantifying expert consensus against the existence of a secret, large-scale atmospheric spraying program" [155]. It was published via IOP Science in a journal called *Environmental Research Letters*.

Looking at the list of four authors of this paper, there is one peculiarity. Three of the authors (Christine Shearer, Ken Caldeira and Steven J Davis) are associated with one or more universities. The fourth, Mick West, is associated only with his metabunk.org website, and not even with his (surely more relevant) "contrail science" website.

The letter is simply a presentation of the opinions of 77 "experts" on a small portion of the evidence that is discussed here. There is only one photograph shown in this study – that of single, broken trail.

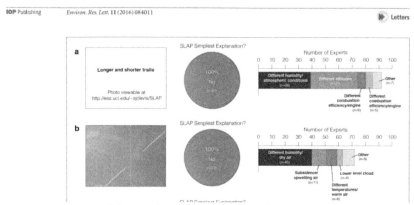

Figure 66 – One of the graphics from the "scientific" letter by Mick West et al.

There is no discussion of grids, circles or other trail anomalies. However, we see, for example that this paper is discussed in a posting on a site called "Science Alert" [156] – where the headline image shows an enormous number of overlapping trails, thus:

Figure 67 – Headline image from an article about the scientific paper, dismissing the PJT issue.

The "letter" (study) is not really scientific in terms of examining *empirical evidence* related to the trails themselves – it merely *collates responses to questions about them*. Even so, the letter/study actually gets a reference from the prestigious *New Scientist* website [157]. Perhaps unsurprisingly, it is in a reposting of an article [158] written by "Bad Astronomer" Phil Plait [159], though that posting has a different title. Why would an astronomer be "called in" to "debunk" the idea of aircraft trail anomalies? Why would *New Scientist* publish his article when he's not a climatologist, meteorologist or any kind of specialist in aspects of aircraft engine operation, aviation fuel or a similar topic?

An Acknowledgement of the Effects of Trails?

From time to time, articles "pop up" which seem to acknowledge the effect of the ongoing trailing. One example is an article on Smithsonian.com from 16 December 2015 [160].

> *If you go outside on a clear day and look up toward the sun—being careful to block out the bright disk with your thumb—you might see a hazy white region surrounding our star. This haze is caused by airplanes, and it is gradually whitening blue skies, says Charles Long of NOAA's Earth System Research Laboratory in Boulder, Colorado. "We might be actually conducting some unintentional geoengineering here," Long said at a press conference this week at the American Geophysical Union Fall Meeting in San Francisco.*

This article tells us very little about the phenomenon and does not even reference the Dr David Travis study of what was dubbed the "Global Dimming" related to contrails [161]. It continues (emphasis added):

> *Long and his colleagues **don't yet have enough data to know** how much of an effect the icy haze left by airplanes may be having on the climate or whether it is contributing to warming or cooling. But its existence demonstrates yet another way that humans might be altering the climate system, Long says, and "you can see this with your own eyes."*

They also seem to suggest that they don't know why the sky is clearer on some days than others:

> *Skies aren't clear all the time, and why they are clear one day and not the next could matter. "The reason for the clear sky is a factor [Long] needs to explore more," Trenbeth says.*

I suggest that this is a subliminal message saying "give scientists more money to investigate, but not find an answer" …

Questions and Petitions to Official Bodies

There has been little, if any, formal investigation into PJT phenomena, though a number of people have tried to raise questions formally using an FOIA[162] and through lobbying politicians – such as through the Skyguards group[163], which organized a meeting in the European Parliament in

Brussels in April 2013 [164]. In 2007, Rosalind Peterson gave an address to a Climate Change Conference, organised by the UN, in New York [165].

Many hundreds or even thousands of "YouTubers" have uploaded videos of various kinds – some are particularly strange and show planes trailing together – with examples in Germany[166] and in the USA[167]. A number of independent, good quality documentary films have been made by people such as Patrick Pasin[168], Clifford Carnicom[169], and Michael Murphy[170]. There are others of varying quality.

Meeting at European Parliament in Brussels – 08 and 09 April 2013

In early 2013, I was invited to speak at this conference in the EU Parliament in Brussels. The conference was organised by Josefina Fraile Martin, from Spain and included speakers from Greece, France, Belgium, Ukraine, Germany and the USA. The conference was supported by MEP Mrs. Tatjana Zdanoka and by Italian Journalist and former MEP Giulietto Chiesa.

On 08 April 2013, in the evening, Michael Murphy's film *Why in the World Are They Spraying?* was shown to an audience of 40 or 50 people – each of whom had to register prior to the event. On 09 April 2013, the conference was held in the same room in the EU Parliament and attendees/delegates also had to register before the conference. About 60 people attended but there was only one MEP, Mrs. Tatjana Zdanoka. For the question and answer session, we had to move to another room, but we only had 30 minutes. This conference and Q & A session were filmed and are available online[171], along with other materials that were discussed at the conference. These same materials were given to conference delegates on a DVD containing various files – a link to which can also be found below. There was little follow up to this conference and though a press release was written[172], it did not seem to have been widely circulated. (I certainly didn't receive any questions about it.)

On 09 April 2013, I had chance to talk with Michael Murphy, who had created the *What in The World Are They Spraying?* film before the *Why in the World* Are They Spraying. Though I have since become much more sceptical of the "spraying" explanations, his films do have some useful information in them, such as exposing the sorts of ideas that are being discussed in consultation meetings by people such as David Keith and Ken Caldeira. Again, I have found that Michael Murphy has not discussed the evidence I cover later in this book, in relation to weather modification. (I have often observed that those discussing "chemtrails" won't talk about certain topics.)

Other People Who Have Tried to Raise Awareness of Anomalous Trails

There are probably too many others to list, but two people that I have noticed are Max Bliss[173] and Dane Wigington (of GeoEngineeringWatch.org)[174] – who have seemingly expended considerable time and energy in challenging certain people over the trailing issue. Though I agree that the "official truths" need to be challenged, these two fellows in particular will, yet again, not accurately discuss what is in parts of this book, preferring to promote the "toxic spray" ideas (almost exclusively). Dane Wigington[175] often promotes a form of catastrophism of his own – for example saying that global warming could be a "runaway issue" because of methane emissions, not CO_2 emissions.

Lack of Genuine Whistleblowers

In relation to any active or ongoing global spraying programme – using any kind of aerosols – no genuine, knowledgeable whistleblowers seem to have come forward with detailed information that can be supported by comprehensive documents, photographs or videos. Though there has been ongoing internet chatter since about 2005 about people like A.C. Griffiths [176] and Kirsten Meghan [177], they have not presented any verifiable, solid information[178]. Though they may have referred to documents such as *Owning the Weather by 2025* [179] (discussed later) and other documents that have been produced by the military, they have not explained many – or even any – of the observations we have made.

Summary

The trailing phenomenon is real. No scientific explanation of all the strange effects that have been photographed and video recorded has been forthcoming. Some researchers seem to promote an explanation of the phenomenon and exclude all other explanations and evidence which would prove their explanation wrong. "Whistleblowers" that have come forward have added little or nothing to our knowledge of the cause of the phenomenon (i.e. "those that know don't talk and those that talk don't know.)

I would argue, then, that we have another dialectic here – with academia saying "there is no cover up – we are seeing ordinary contrails." In opposition, "chemtrail" groups and websites insist we are being poisoned, daily, with a toxic spray. The truth is that "something else" is going on. We will discuss what this might be in Chapter Twelve.

CHAPTER EIGHT
ARE CONTRAIL PATTERNS CAUSED BY COMMERCIAL AIRLINE FLIGHTS?

"To treat your facts with imagination is one thing, to imagine your facts is another."

John Burroughs (1837–1921)

Space for Notes Below

Introduction

In this chapter, I describe attempts to track and record aircraft movements and compare these movements with observable trails left in the sky. The aim was to try and prove a correlation between the movements of identifiable aircraft with the appearance of persistent jet trails.

Useful Information

As described earlier in this book, the petitions to all official bodies have yielded little useful information and exactly no information which can explain the formation of grids, circles and multiple intersecting trails. Official sources did not even give me information about the Aircraft Signalling System known as ADS-B, nor the aircraft tracking websites that exist.

ADS-B – What is it?

Many aircraft are now transmitting ADS-B (Automatic Dependent Surveillance – Broadcast)[180] messages when in flight. These messages contain the following information

- A code number identifying the aircraft (sometimes called "ICAO")
- Flight Number
- Altitude
- Position (Latitude/Longitude)
- Speed
- Heading

ADS-B Receiver/Decoder

In about 2006, I became aware of a device that could be purchased which would allow the tracking of aircraft if an aerial was positioned in the right place. The device was plugged in to a PC and software on the PC was run to show on a graphical display where the aircraft were located. It was called a "Virtual Radar." However, the unit was expensive, costing about £500.

Figure 68 – SBS-1   Figure 69 – AirNav Radarbox

It wasn't until four years later, that at some expense, I purchased an AirNav Radar Box to enable further quantification and identification of Air Traffic, in relation to studying trail persistence/non-persistence.

Detecting Aircraft Flying Over Your Location

In the last few years, websites such as www.FlightRadar24.com[181], Flight Aware[182] and Virtual Radar[183] have offered tracking and aircraft identification features, though they can be slow to update and somewhat cumbersome to use. Similarly, there are Android and iPhone Apps[184] which interface to these online services and allow you, for example, to identify flights by holding up your phone in the direction of a plane in the sky. Of course, not everyone has an iPhone or Android phone ...

Figure 70 – Aircraft tracking App or Webpage running on an iPhone

As far as I am aware, the website and phone app solutions don't have logging features of any great sophistication, so are not much use other than for "real-time viewing and tracking."

Planefinder.net does have an "archive" feature, but I wanted something that would be able to *count* the number of planes detected, so I could run some analysis.

AirNav Radarbox

In 2010, I decided to invest in an AirNav Radar Box as I was still very curious as to what could be determined from using one to track aircraft. An important feature was that of "logging" any aircraft it detected – this meant that the unit could be left unattended and data could be examined retrospectively. However, there was still no easy way to get a visual record of trailing, other than deciding to go out with a camera and photograph the sky during periods of trailing. This was not very practical, as I could not devote the required amount of time exclusively to a "tracking project."

AirNav Software

The software that was shipped with the Radarbox provided a "virtual Air Traffic Controller's (ATC) display" – all quite natty, but its logging features were limited. For example, it could keep a list of all the aircraft detected – and it could even playback a recording of logged data, but it was not able to produce charts or, for example, count the number of aircraft detected during a specified period, such as 30 minutes.

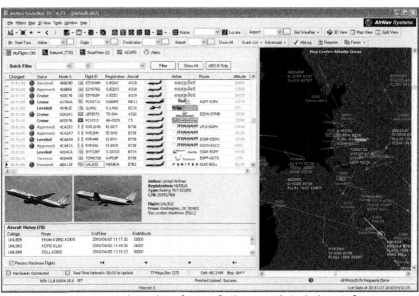

Figure 71 – Typical Display of Aircraft Shown with Radarbox Software

The aerial was positioned in the attic of my house. Though this arrangement did not give the maximum detection range (an aerial positioned as high as possible, outside the house obviously gives the best detection range), it was more than suitable for this project, which was concerned with photographing aircraft as they travelled *over* the house – a range of about 20 miles was adequate.

Though this software possessed logging features (i.e. the captured aircraft data could be saved to a file), if the software was left running for, say, 24 hours, it would create large files and sometimes the software would just "freeze" and so data would be lost.

Easyjet "Long Trail and Short Trail" Data

Having run the tracker for several weeks, I was able to capture some data and then was able to pick up two Easyjet Flights on the same day. I was also able to photograph the aircraft with a Canon SX20 (which had a 20× Optical Zoom. These photographs were taken with a Canon SX20-IS which is a 12 Megapixel unit and has a 20× optical zoom. The lens system employs an optical image stabiliser. (There was approximately a one minute difference between camera time setting and PC time setting).

This incident took place on the date shown and indicated an unexplained difference in the trail caused by two identical planes, as shown below. One plane was seen to leave a very long trail (which I videoed as it was too long to see clearly in a single photo) and the other had a short trail. The details obtained from the software are shown below.

Date: 05 March 2010

Time	Flight	Reg.	Route	Aircraft
16:32	EZY067	G-EZIR	Luton – Glasgow	A319/111
17:24	EZY239	G-EZIN	Stansted – Edinburgh	A319/111

Time	Flight	Min Ht	Max Ht	Speed (Knots)
16:32	EZY067	32525	33675	380-390
17:24	EZY239	32900	33000	384

Are Contrail Patterns Caused by Commercial Airline Flights?

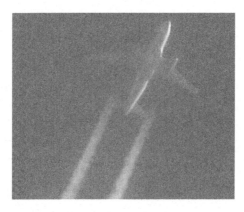

Figure 72 - 16:32 - EZY067 - taken facing roughly west.

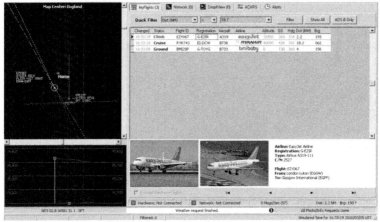

Figure 73 - AirNav Radar Box Track - EZY067

Are Contrail Patterns Caused by Commercial Airline Flights?

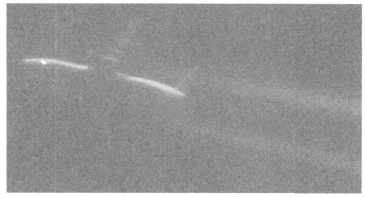

Figure 74 – 17:24 – EZY 239

Figure 75 – 17:24 – EZY 239 – Photo taken facing roughly east north east.

Figure 76 – AirNav Radar Box Track – EZY 239

113

Photos of Trails Left by These Two Aircraft

Figure 77 – 16:32 – EZY067 – Long Trail

Figure 78 – 17:24 – EZY 239 – Short Trail (why does only one trail persist here?)

It was pretty clear that this was an anomaly and so I wrote to Easyjet in June 2010. They never responded to the letter sent by post, but they responded thus (in 48 hours) to an e-mail I sent. This response is reproduced below:

> From: easyJet Customer Experience Team [mailto:easyjet@maileu.custhelp.com]
> Sent: 14 July 2010 12:21
> To: ad.johnson@ntlworld.com
> Subject: Letter Sent by Post on 04 June – no response [Incident: 100713-020119]
> Recently you requested personal assistance from our on-line support centre.
> Subject Letter Sent by Post o 04 June – no response
>
> Discussion Thread Response (Jacqueline Kenny) 14/07/2010 11.21 AM
> Dear Mr Johnson,
>
> Thank you for contacting us.
>
> I apologise but I am unable to answer your operational query.
>
> I would like to inform you that if you wish to talk to one of our representatives, our Customer Experience Team is available under the following telephone numbers: In the UK, 0871 244 2366 (calls cost 10p per minute; calls from mobiles and other networks may cost more).
>
> Our Operations department is option 1. If you are calling from abroad, please telephone the relevant number below ...

Nine-Month Air Traffic Investigation using Raspberry Pi Network System

Due to my concern over the trailing issue, I continued to look for a way to study air traffic movements over my house. The end results did not appear to prove there is a conspiracy to spray aerosol compounds in the sky – even though it is possible that this is actually happening, at certain times and in certain places. It was simply an attempt to try and match or collect air traffic counts and log aircraft movements and then correlate this data with the appearance of trails. In this regard, at least, it has served a useful purpose.

Raspberry Pi.

The Raspberry Pi is a small, credit card sized, fairly powerful computer released in 2012 [185], which runs a version of an Operating System called Debian Linux. I had obtained one not long after the release and set it up as a low-powered file server.

Figure 79 - Raspberry Pi - Model B Single Board Computer

It is a credit to the way that Open-Source software systems work that allows developers now to plug together software and hardware components and build both hobbyist and professional projects – to a high level of sophistication – relatively quickly. With appropriate programming knowledge, customisation of software is straightforward and practical. Coupled with the vast and easily searchable resources on the internet, solutions to common problems can quickly be found, enabling system reliability to be improved much more readily. Significant computing power in a small, cheap and energy-efficient package also means that more and more advanced projects can be envisioned and developed at a modest cost of only a few hundred pounds.

In June 2013, I wondered if it was possible to connect the Raspberry Pi to the AirNav Radar Box – essentially to replace the Netbook and allow the Pi to take the data from the AirNav box and save it, so that I did not have to tie up a Netbook for this purpose.

After finding a forum discussion about this, I also found another and potentially better way of doing a similar sort of thing, and "PiTracker" started to become a workable idea.

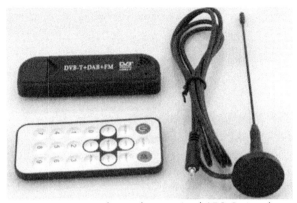

Figure 80 – Dongle used to received ADS-B signals.

A page by Dave Taylor[186] provided a solid basis for some further Raspberry Pi development. By getting the right type of USB Dongle – a Digital Terrestrial Broadcast Receiver Dongle (DVB-T) with the correct chipset (R820T/RTL2832U), I could track aircraft in real-time using a Raspberry Pi. Hence, all that was now needed was additional software to do the logging and counting. This was made much easier because there was a program, called Dump1090, which decoded the ADS-B messages and also presented data from them through a web page interface. This program was written in C. In other words, all the hard work of decoding ADS-B messages was already done – I just needed to add some code to count the detected number of aircraft and generate charts.

Counting Aircraft

It was relatively straightforward to adapt the Dump1090 program code to make it count detected aircraft in a set period. It was also possible to get it to count aircraft in various categories – such as those above 25,000 feet, where trails are formed. All these counts were saved into a "daily data file." Additionally, a log of all aircraft detected was generated and saved. The main software development was done using an Ubuntu Linux installation in an Integrated Development Environment (IDE) called CodeBlocks. (The TV Dongle and Dump1090 code was modified on a PC running Linux and a method of creating "tracking charts" was developed.) The C code was simply copied onto the Raspberry Pi and compiled so that it would run on the Raspberry Pi directly.

Photographing the Sky

In May 2013, a custom camera board was released for the Raspberry Pi and this could be operated by software that ran on the Raspberry Pi. It was now therefore possible to have the Pi log and track the aircraft – and photograph the sky – unattended, and using less than 8 watts of power.

Additionally, Raspberry Pi camera images were of considerably better quality than the Web Cams, as the Pi Camera has a 5-megapixel sensor.

Figure 81 – Raspberry Pi – 5 megapixel camera board

Automatic Capture of Weather Data

Using the World Weather Online website – www.worldweatheronline.com – it was possible to obtain weather data at regular intervals, to be saved with the air traffic counts. Admittedly, ground-level weather data is not especially useful in relation to conditions which may affect the formation of trails at 25,000 feet and above.

Configuration Data

In order to generate meaningful data, it was necessary to add a "configuration feature." Most importantly, the latitude and longitude that the Raspberry Pi was located at needed to be set up – this would then allow measurements to be made (mainly the plane's distance from the Raspberry Pi's location).

Webserver/Webpage to Display Real-time Plane Positions

The Dump1090 software also generated a webpage which showed the positions of detected aircraft on a Google Map in real-time, along with any available data about each detected aircraft. However, this webpage view defaulted to show a location near London, so this part of the software was also modified to display a map based on the configured location. Additionally, the webpage was modified to include additional features, such as aircraft counts and local weather data.

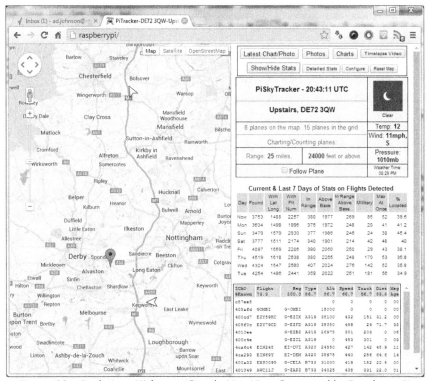

Figure 82 - Real-time-Webpage/Google Map View Generated by Raspberry Pi Tracker

Multiple Trackers – Remote Configuration and Upload of Captured Data

In order to get a better sample of data, it was decided early on that several "PiTrackers" should be put into operation, so several volunteers from around the UK were asked to host them at their homes. I am grateful to those five volunteers who agreed to host trackers and help me set them up. Without their help, this project would not have been able to gather nearly as much data.

This meant that a method had to be developed for transferring the data captured by these trackers to a central location (my own Raspberry Pi file server!) Hence, existing scripts were modified and a server was configured to accept and store the uploaded time-lapse video and aircraft data. Additionally, working with volunteers, the trackers were, when possible, set up to be remotely configurable. This allowed software to be modified and updated (as was necessary on more than one occasion). I could make changes to the tracking software, or one of several Linux scripts, to keep things running as required.

System Components and Overall Operation

This photograph below shows the main components of the tracker system. The system uses an unmodified Raspberry Pi with an SD Memory card (like those used in Digital Cameras and similar devices). The memory card holds both the Raspberry Pi Linux Operating system ("Raspbian OS") and it is used to store the data acquired from the aircraft, as well as photos taken by the Raspberry Pi Camera.

Trackers were placed, when possible, on an upstairs window sill, which had a clear view of the sky. Once configured with a postcode, latitude, longitude and station name, they were left running 24 hours per day, seven days per week. The tracker software included features to calculate local sunrise/sunset times and would only capture images and create plane charts during local day time.

Linux "scripts" and commands were created to compress ("zip") each day's data files and upload them to the server between midnight and 6 am. Similarly, time-lapse videos were generated and uploaded to the server every night.

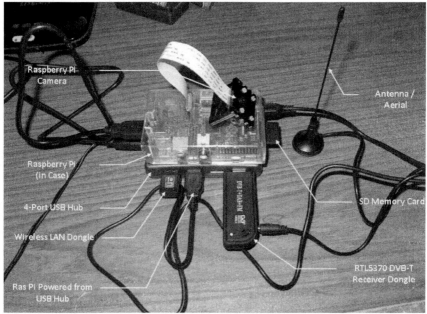

Figure 83 – Single PiTracker in operation.

Plane Charting

By using Linux Open Source Graphics Libraries (libplot and libglib), it was possible to plot aircraft paths on charts – as the data was captured by the Raspberry Pi. Charting parameters could be set so that planes within a certain range were drawn on the charts (which were created every 30 minutes by default). Only planes above a certain (configured) altitude were logged on the chart.

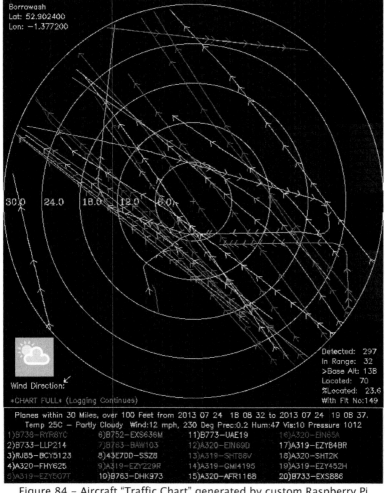

Figure 84 - Aircraft "Traffic Chart" generated by custom Raspberry Pi Software (100-foot base altitude)

These charts were saved in PNG format (a useful feature of the graphics libraries). Aircraft tracking data was saved in CSV Format. This data was ultimately imported into a Microsoft Access database for analysis.

Time-lapse Videos

The trackers were set to only take images from half an hour before sunrise to half an hour after sunset. After realizing that the sky needed to be photographed approximately once every minute, it was realized that several hundred photos per day would be generated and these would need to be reviewed to check for trails. Clicking through hundreds of photos per day would have been a slow process, even using something like Google Picasa – which has a very fast/responsive image viewer, so it was soon determined that the Raspberry Pi was capable of automatically generating time-lapse video files (in MP4 format) by using another package called libav-tools.

The original intention was then to review photos taken by each tracker and count any trails that appeared in each image, noting the time of the appearance of trails, when a particular trail or trails appeared on more than two consecutive images.

Tracker Reliability and Continuous Operation

Over a period of several weeks, the reliability of the tracker was tested – could it run autonomously for days or weeks at a time? Updates to the Raspbian OS were still appearing every few weeks and by about September 2013, it seemed that trackers could run for long enough periods. However, it was still necessary to implement strategies to ensure continuous/smooth running of a tracker. A 4GB memory card was used on some trackers and an 8GB card was used on others. In both cases, however, it was not exactly clear how many days it would be before the card became filled with saved data. Most storage space was used by the time-lapse videos and the hundreds of camera images. In practice, it turned out that between about 15 and 25 days' worth of data could be stored. A method was therefore added to delete data after a certain number of days (this could normally be done safely, because each day's data was uploaded to the server every night – and the upload method proved fairly reliable).

Cloudy Hours and Days – Sky Blueness

As time-lapse videos were reviewed, another problem became apparent – in the UK, where all the trackers are situated, days where the sky is completely free of cloud for more than a few hours are rare. A whole day could be reviewed and there were no usable sky images at all – as the weather was too cloudy to observe trails. A method had to be devised, therefore, to measure how "clear" (blue) the sky was – which allowed a determination of how usable the image was. A software package called "Image-magick" (which was used to add timestamp text to images and make a composite image with the plane chart and weather icon) was therefore used to analyse each sky image, immediately after it was captured, to determine the "blueness." This was calculated as a number which ranged from about -85 to +250. Following some tests, it was determined that in most cases, images that had "blueness values" of greater than about 50 could be used to look for trails.

Microsoft Access Database

Daily Data (CSV) files from each of the trackers were merged together every few days or weeks and imported into a "holding" table. Database (SQL) Queries were developed to sort and group data by location, date and time. This data could then be presented on forms for inspection and modification. For example, a "notes" field was included so that any unusual trails or weather effects could be noted while sky images were reviewed.

Counting Trails – Entering Data into the Database

Once data was imported into the database, a query could be run to determine if there were any periods in each day when the sky was clear enough for trails to be observed. A database form was developed which allowed trail counts to be entered for the times that the sky was clear enough to see them. i.e. the query would "filter out" any days where there were no periods clear enough to see trails, which saved some time during reviewing time-lapse video. It was also noticed that the "blueness factor" was generally a fairly good indication of when trails would be visible, although for example, if the Pi camera was pointing at the sun, and there was some haze in the sky, the image would not appear to have much blueness, but it was still possible to see some trails. So, the blueness factor calculation was not always reliable. (Data between about September and November 2013 was inspected by volunteers and trail counts were added by them – this data was imported into the database too).

The time-lapse video was reviewed in VLC Player – which was used to step through one frame at a time (by pressing the E key). This allowed closer examination of some video frames for some sections of each video.

Counting Detected Aircraft

When the software detected a "new" aircraft, it was counted. Once it had been detected, it was assumed it could remain in range for 90 seconds and therefore it would not be "re-counted" if the tracker temporarily lost the signal from the aircraft (for less than 90 seconds). If the same aircraft was detected by the tracker at periods greater than 90 seconds apart, it would be counted twice.

Aircraft Broadcasting Their Location

Even when originally using the AirNav Radar Box, it was discovered that only a certain proportion of aircraft seemed to be broadcasting their location data (latitude/longitude). Nearly all would broadcast their altitude and call sign.

Military Flights

These were identified based on data in the database which was packaged with the AirNav software. It seemed none of these flights broadcast their location. However, it was at least possible to count them when they were in range.

Aircraft Altitude and Range

The Tracker software was designed to allow aircraft in certain altitude bands to be counted and these counts were stored separately. Total counts of aircraft detected above the base altitude were stored. (It was assumed that aircraft at or above 25,000 feet had the potential to form trails and aircraft below this altitude should not form trails.)

Depending on where the tracker was sited, it could detect aircraft over 100 miles away. For the purposes of this study, it was considered that an aircraft within 25 miles of the tracker could be seen if it was leaving some type of trail. When the data was analysed, this figure did not seem unreasonable.

"Above Base and In Range"

Hence, most of the counts and figures were considered with regard to the assumption that for a trail to be viewed, an aircraft would have to be in range of the tracker and above the base altitude for forming a trail. A count was therefore kept of aircraft that fulfilled both these criteria.

Main Counts to Be Determined

Taking the factors/assumptions in this section as a whole, the main objective was therefore to try and establish if the following average counts differed on days and/or during periods when trailing was observed compared to days/times when no trailing was observed.

- Count of planes detected
- Count of planes located (those which had latitude and longitude)
- Count of planes above trail altitude

- Count of planes which came within 25 miles of the tracker's location
- Count of planes above trail altitude and within 25 miles of the tracker's location

This also necessitated that the "blueness factor" was used to determine periods of clear skies, so that a fair comparison of figures could be attempted.

Trail Count Considerations – Camera Field of View

The Raspberry Pi camera has a 49-degree horizontal field of view and a 37-degree vertical field of view[187]. This means that it can only capture about one eighth of the horizontal view and about one fifth of the vertical field of view. This therefore means that trails may have appeared on some days, but not been picked up on any time-lapse videos. Conversely, it means that it is possible that on days when light trailing *was* seen, there may have been many more trails than indicated in the counts entered into the database.

General Assumptions

It was generally assumed that air traffic over the locations where trackers were sighted followed a weekly cycle – and it was mainly civilian in nature. That is, the volume of air traffic on every Monday would be roughly the same, and the volume of air traffic on every Tuesday would be roughly the same etc. The trackers were kept in the same physical location during the time they were running – this was important because if they were moved, say, from an upstairs room to a downstairs room, an immediate effect on the ability of the trackers to detect aircraft could be observed. Also, with repeated observation, it could be seen that a tracker may detect aircraft at a greater range in one direction (e.g. looking towards the southwest). This was likely to be a result of a clear line-of-sight view to the horizon in this direction. That is, a signal from the aircraft is impeded by buildings, trees etc in between the tracker and the aircraft. However, this should be less of an issue for aircraft that came into the range where trails were to be observed, because the signal would be stronger when the aircraft was nearer to the tracker.)

Daily Count Totals

It was necessary to develop an easy method to sum all the required counts from all the half-hourly periods for which the tracker logged counts – this was done using SQL queries to collate results.

Example Images and Charts

The images and charts below were used to identify some of the flights. On the charts, the centre of the chart represents the tracker's location (which was configured using a latitude and longitude value obtained from online maps). North is at the top. Using these two pieces of information, it was almost always possible to identify any flights which were successfully tracked.

Figure 85 – Lower trail was likely created by flight EXS718N

Are Contrail Patterns Caused by Commercial Airline Flights?

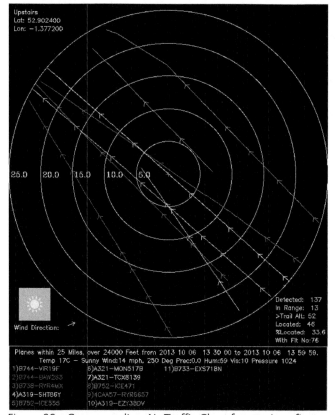

Figure 86 – Corresponding Air Traffic Chart for previous figure.

Figure 87 – Trail was likely created by BAW3304

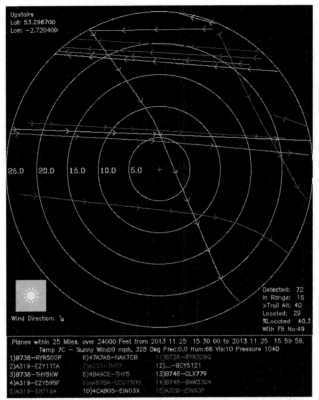

Figure 88 – Corresponding Air Traffic Chart for previous figure.

Figure 89 – Curved Trail was likely created by flight VIR6J

Are Contrail Patterns Caused by Commercial Airline Flights?

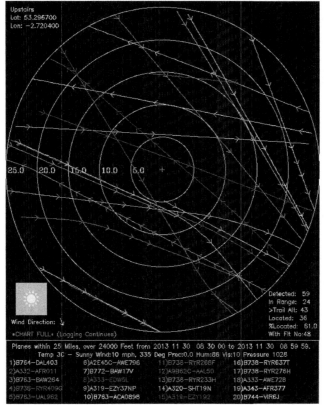

Figure 90 – Corresponding Air Traffic Chart for previous figure.

Contrail/Distrail

On the Leeds tracker – 08:30 am, 04 December 2013 – two of these curious trails were observed in quick succession and these are less commonly seen than the "ordinary" trail. There were also unusual "shadow" effects. If you look carefully at the image below, you can see where the trail should be but it isn't there – there is a kind of shadow, however. In some cases, it seems difficult to accept that the shadow has been caused by the sun shining on visible parts of the trail…

Figure 91 - Distrail over Leeds 04 December 2013

Figure 92 - Distrail and a shadow over Leeds 04 December 2013

Figure 93 - Distrail and a shadow over Leeds 04 December 2013

Detected Air Traffic Levels and Trailing

From all the data gathered over nine months, it seems there is no large difference in "ADS-B detectable" aircraft on days of high trailing than there are on days of no trailing. The data here does *not* establish a clear link between levels of aircraft and levels of trailing. Looking at some figures in isolation, it could be argued that there is a *lower* number of aircraft on days of heavy trailing.

Perhaps a better detector is needed, although this is unlikely – as a range of 20 miles should be sufficient to "detect planes and see trails" with this sort of equipment. For example, if one examines the charts, most of them have an unbroken line of travel for the planes, which means enough of the messages were picked up, while the plane was in range, to plot the path of the plane.

Potential for Grids to Form

In this chart from the WA6 tracker, it is relatively easy to see the potential for grids of trails to form:

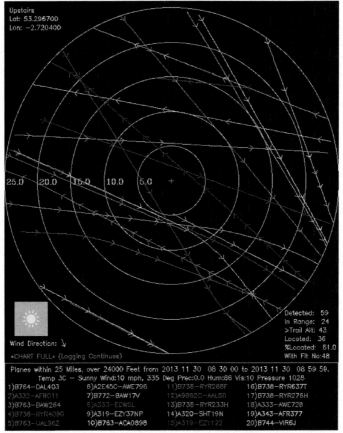

Figure 94 - Air Traffic Chart during grid formation

However, further examination of the captured/generated charts and time-lapse video was needed to establish if this could actually be seen happening.

In the 25 November 2013 time-lapse video from the WA6 tracker, we can actually see two grids form, at approximately 7:53am and 14:36pm:

Figure 95 – Contrast enhanced image to show grid of trails

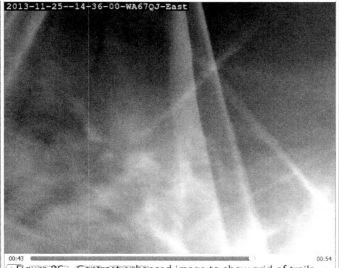

Figure 96 – Contrast enhanced image to show grid of trails – (central vertical lines are a reflection of the curtain)

I studied the 54 second video carefully (consisting of approximately 540 frames) and was interested to see when planes were travelling something near to a north/south direction. This happened at the approximate following times: 07:39, 07:43, 07:48, 07:57, 09:35, 10:20*, 10:30*, 11:09*, 11:34, 11:45, 11:59, 12:18, 14:12, 14:23, 14:36. Therefore, about 15 planes in all were travelling north to south, or in a path which was close to this direction. The ones marked with an asterisk did not leave a persistent trail, only a "normal" one. These flights seemed to be travelling directly over the house – appearing fairly centrally in the picture. Thus, it should have been easy to see these flights plotted on the charts generated for the appropriate 30-minute period.

Studying the 19 charts generated on 25 November 2013, only two possible flights seemed to be travelling in the correct direction. The 11:09 was probably SHT 7W (no persistent trail). The only other flight logged travelling north to south or north-west to south-east (over the tracker's location) was at about 3:15pm (see chart below).

Figure 97 – Trail central and just above sunburst is likely SHT7W

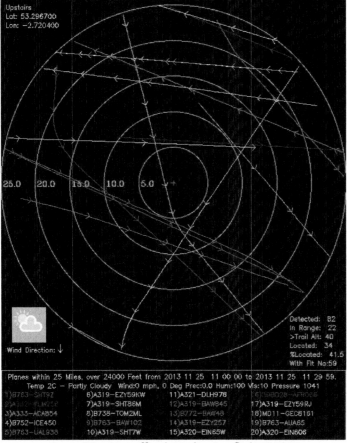

Figure 98 – Air traffic chart showing flight SHT7W

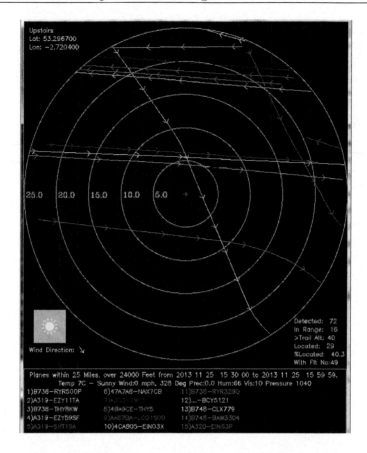

Figure 99 – Flight (4) A319-EZY59F was the only other flight shown on all the 25 November 2013 charts which was travelling in a NE – SW direction over the tracker (at the centre of the chart)

So, I would really like to see someone document trail grid formation and document *all* the flight numbers and types of aircraft that made them. I have invested considerable time and effort in doing this and, according to "mainstream" sources, it should be easy and obvious. What I have shown is that it is neither.

Identification of Flights Leaving Trails and Formation of Grids

It was sometimes quite difficult to identify which flights left trails – perhaps because only a maximum of about 50% of the flights could have their latitude and longitude decoded. Without this information, it was impossible to know if the flight was right overhead, or whether it was 100 miles away. The "percentage located" figure showed no appreciable variation between days of trailing and no trailing.

The study from 25 November 2013 of the WA6 tracker shows that it was not possible, using ADS-B data, to identify all of the flights which made the grids. Though, looking at some of the charts, the potential for grids to form can be seen, these were not seen "in the right place" and not enough flights were detected to prove, from this data, that civilian air traffic is responsible for forming these grids. So, they remain a mystery.

A study of data from the DE72 tracker seemed also to show that flights that were travelling in a north-south direction were rarely seen on charts – and if they were, it was over the far west of Derby – about 20 miles from where the tracker was sighted. Further attempts at identifying "trailing flights" could be undertaken, even with the existing data. It also seemed hard to identify flights travelling north-south on the WA6 tracker.

Days of Trails or No Trails?

No obvious reason could be observed why trails were seen on some days and not others. Again, taking the 25 November 2013 time-lapse video, it can easily be seen that there are persistent and non-persistent trails appearing in the same periods of time and the same part of the sky.

So, this remains as one of the principal unanswered questions. Why do we have days when no trails appear – not even ones that persist enough to actually see them; and then on other days, we can observe many, many trails for such a length of time that they can even seem to spread out and form a "haze blanket"?[188] There seems to be no satisfactory explanation for these different scenarios, beyond either "hand waving" or making claims which are not supported by the evidence. For example, if it is caused by the state of the jet stream, and its influence on the stratosphere, there has been no clear explanation as to exactly what sort of circumstances/conditions would cause trails to persist for many minutes and, specifically, how jet stream changes would cause these conditions to change. Some would claim that papers like the one written by Dr Ulrich Schumann provide some kind of explanation – but this is untrue. This particular paper did not contain any time lapse studies, for example.

CHAPTER NINE
PERCEPTION MANAGEMENT OF THE PJT/CHEMTRAIL ISSUE

"Nothing doth more hurt in a state than
That cunning men pass for wise."

Francis Bacon

Space for Notes Below

"Nothing to See Here!"

Let us suppose, for a moment, that something is happening above us which the people that are running the planet don't want us to notice. Some people think it is a large scale toxic spraying programme of some kind. Some think it could be to make us sick or to slow down or stop our evolution "onto a higher level of consciousness."

However, I myself don't quite know what the reason is, but I am strongly convinced that we are indeed being discouraged from paying attention to what is happening in our skies.

Here, I will show some evidence of what seems to be deliberate *perception management* relating to the appearance of the persistent jet trails (whatever name you give them).

Chemtrails/Jet Trails in Advertising and TV Visuals

There does seem to be an unusual prevalence of jet trails in advertising and in places where you might not expect them to be shown – I have collected some examples on this page[189].

Virgin Trains/The Railway Children

One especially curious example was seen in a 2005 Virgin Trains commercial [190]. Here, some original 1970 film footage was digitally composited with newer images – including an aircraft trail – as shown below.

Figure 100 – Railway Children – advert and original film comparison.

When I showed this to another researcher called Phil Morris, he took it upon himself to find out who made the commercial. He then wrote to them and received the response below.

> *From: xxx*
> *Sent: 19 July 2010 09:53*
> *To: PHIL Morris;yyy; ad.johnson@ntlworld.com*
> *Subject: RE: The Railway Children, Virgin Trains and Chemtrails*
>
> *Hi Phil*
>
> *The chemtrails were indeed 'on-purpose' in our Virgin Trains 'Return of the Train' ad.*
>
> *If you look closely in the background you can also see a block of modern flats on the horizon. This juxtaposition of old and new aims to highlight the new beginnings of the modern Virgin Train.*
>
> *In 2004, Virgin Trains began to roll out their new fleet of new trains, the Pendolino, to the West Coast line. It was deemed to be time to address their ultimate ambition – to become the nation's transport provider of choice. We wanted to encourage*

> consumers to make an active, positive choice to take the train rather than driving or opting to fly – and to keep doing so over time.
>
> So, as you can see, the chemtrails were used as a tool, although subtle, to address the above.
>
> Many thanks

Level 42 – Lessons in Love – 1986

If you watch the music video all the way through, you see a large number of trails are displayed as a backdrop as the group plays the song.

Figure 101 – Still from Level 42's "Lessons in Love" video

Subliminal "Chemtrails" – in kids' movies and on a book cover

The following examples are from CGI films – hence the trails in the sky were added in deliberately and don't form part of the story line, as far as I can tell. I guess there are quite a few more I haven't included ....

Figure 102 – Over the Hedge – Dreamworks (2006) [191]

Figure 103 - Over the Hedge – Dreamworks (2006) Error! Bookmark not defined.

Figure 104 – Cars – Disney/Pixar[192] (Contrast Enhanced for clarity)

Figure 105 – It's a war-time story, but why the grid?[193]

Advertising/Billboards etc

Figure 106 – Spirit Aerosystems[194] – Date Unknown – but approximately 2007

Figure 107 – Talk Talk UK Broadband Service – June/July 2012 [195]

Further examples can be found online.

BBC Wimbledon Highlights – Opening Titles – 2013 – 2015

If you take one or two minutes to watch a video of the titles, you should note that *every single shot of the sky* shows one or more trails. I don't know about you, but as this is primarily a CGI video, I'd prefer to see fluffy clouds, not pollution ... Was it just re-use of stock images?

Figure 108 - BBC Wimbledon Highlights - Opening Titles - 2013 - 2015

Figure 109 - BBC Wimbledon Highlights - Opening Titles - 2013 - 2015

Figure 110 - BBC Wimbledon Highlights - Opening Titles - 2013 - 2015

"Chemtrails are Coal Fly Ash" – Dr J Marvin Herndon

In August 2015, Dr J Marvin Herndon had a paper published in *International Journal of Environmental Research and Public Health*,[196] and people who are interested and/or deeply concerned about the trailing issue seemed to suggest that finally, someone had explained the phenomenon. The paper discusses and observes many of the same factors relating to aircraft trails discussed on this book (and elsewhere) ...

> To date this new threat, posed by widespread, intentional tropospheric aerosol particulate emplacement, has gone unremarked in the scientific literature for more than one decade. Here, based upon original research, the author discloses substantial evidence as to the identification and nature of the specific particulate substance involved and begins to describe the extent of this global public health and environmental threat.
>
> Recently there have been calls in both the popular and scientific press to begin discussions about the possibility of engaging in future stratospheric geoengineering experiments to counter global warming [2,3].

He continues:

> In the spring of 2014, the author began to notice tanker-jets quite often producing white trails across the cloudless blue sky over San Diego, California. The aerosol spraying that was happening with increasing frequency was a relatively new phenomenon there. The dry warm air above San Diego is not conducive to the formation of jet contrails, which are ice condensate. By November 2014 the tanker-jets were busy every day crisscrossing the sky spraying their aerial graffiti. In a matter of minutes, the aerosol trails exiting the tanker-jets would start to diffuse, eventually forming cirrus-like clouds that further diffuse to form a white haze that scattered sunlight, often occluding or dimming the sun. Aerosol spraying was occasionally so intense as to make the otherwise cloudless blue sky overcast, some areas of sky turning brownish.

However, when I read the paper, I realized that it didn't really explain the phenomenon. Herndon writes:

> For three months during a period of intense aerial spraying in 2011, an individual in Los Angeles, California captured and had analyzed outdoor air-borne particulates. The results were posted on the Internet [11]; subsequently the author obtained the analytical laboratory report.

The paper just contains a description of how Herndon compared results of a rainwater sample analysis, taken anonymously in the Los Angeles area in 2011 with a sample of leached coal fly ash which was tested by someone else (Moreno et al). So, the sum total of research in this paper is that Herndon noticed the trailing over his house, went online and found an anonymous rainwater sample analysis of unknown provenance and then compared those results with a test of coal fly ash and concluded from that that fly ash is being dumped out of the sky all around the world! He makes this plain here:

> **The global extent of tropospheric coal fly ash emplacement is inferred** *from rainwater analyses reporting the three elements (aluminum, barium and strontium) that are prominent in the leachate of laboratory coal fly ash water-leach experiments.*

The rest of Herndon's paper simply discusses the toxic effects of chemicals found in the fly ash (and the water sample) and provides no more evidence to prove the source of these toxic substances. Perhaps it is little wonder that the journal then retracted the paper.[197]

In response to this, more mainstream sites wrote things like this [198] :

> *Chemtrails is a pseudoscientific conspiracy theory based on the notion that the government is secretly releasing mind-controlling chemicals that sterilize people from airplanes.*

Whilst criticism of the paper is valid, the statement above is specifically worded to ridicule the phenomenon in a particular way. The criticism appears to refer to Herndon's point, which is probably correct:

> *Aluminum is thought to impair fertility in men [38] and is also implicated in neurological disorders of bees and other creatures [39–41].*

So perhaps there are higher levels of aluminium in the environment, and this has the effect that Herndon and others describe, it's just that the source of that aluminium may not be "spray from planes." For example, it could be aluminium which has come from building materials, cookware or mechanical components in cars that have undergone wear. That is, the effects of aluminium that Herndon describes could be real, but he attributes them to the wrong cause.

David Keith and Geoengineering/Chemtrails

Earlier, we discussed Dr David Keith and his "plan" regarding the atmospheric distribution of aerosol compounds, such as sulphur or "nano-discs" of aluminium oxide and barium titanate[61] to reflect sunlight. Many people who are concerned about these issues were "up in arms" about this[199].

On 09 December 2013 in an episode of the Colbert Report (a light-hearted current affairs programme in the USA), geoengineering was the subject of conversation …

> *Keith: The worst way in which to make decisions about this would be that we all agree we won't talk about it in polite society and we suppress it, which is basically what we have been doing and then ye know in 2030 – suddenly a crisis and we make a fast decision …*

> *Colbert: It's happening already – ever look at those planes up there? They have contrails behind them ... maybe all those contrails – maybe they're actually spraying chemicals in the atmosphere right now and Uncle Sam isn't telling us...*
>
> *Keith: Seems extremely unlikely ....*
>
> *Colbert: ..that the United States is not telling something to its citizens... seems extremely likely to me ...!*

I now put it to you that David Keith and Stephen Colbert are part of a psychological operation to manage people's perceptions of so-called "chemtrails" and geoengineering issues. We see the dialectic being presented in a popular program – so that people who believe in toxic spraying and disbelieve AGW theories and those who advocate the use of geoengineering will be whipped up into arguments.

The Hegelian Dialectic Again ...

So, these examples show how people challenging the mainstream can be rallied around something which is a false explanation for a real phenomenon – then the false explanation is debunked and the original problem (in this case the level and pattern of trailing) is never explained and the real cause of the phenomenon remains covered up. People don't even investigate what is actually happening – they just argue about what they think is happening, without realizing they have made too many assumptions.

CHAPTER TEN
WEATHER ANOMALIES

"Some are weather-wise, some are otherwise."

Benjamin Franklin

Space for Notes Below

Still Nothing to See ...

In this chapter, we will consider photographic evidence of weather anomalies – these photos are both "ground-based" and from satellite images. They represent a tiny sample of those now available. I recommend the reader also studies a number of videos such as those by WeatherWar101 and earlier ones by "Dutchsinse." We will consider explanations which are proposed for some of these anomalies.

Weather Radar Anomalies

A few examples are shown in this section. Again, thousands more are available online at sites such as http://www.radaranomalies.com/.

Australia Anomaly

An article posted on *Adelaide Now* on 1 April 2010 reports:

> The anomalies first began on January 15 when an "iced doughnut" appeared over Kalgoorlie in WA. Satellite imagery showed there was no cloud over the area at the time to explain the unusual phenomenon but farmers' online comments claimed it was "unusually hot" all day.
>
> It was followed by a bizarre red star over Broome on January 22 and a sinister spiral burst over Melbourne described by amateur radar buffs as the Ring Of Fire Fault.

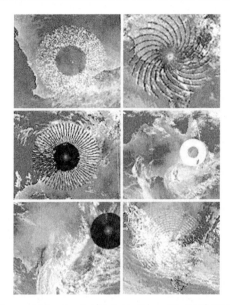

Figure 111 – Australia Anomaly – 01 April 2010

USA Anomalies

This collection[200] is primarily from 1999 and includes animation sequences. What is causing these large-scale circular features?

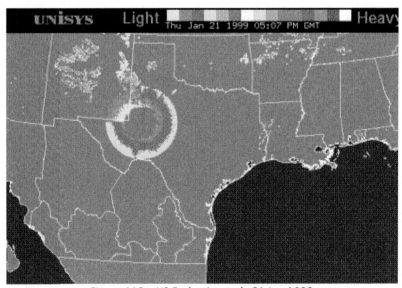

Figure 112 – US Radar Anomaly 21 Jan 1999

Weather Anomalies

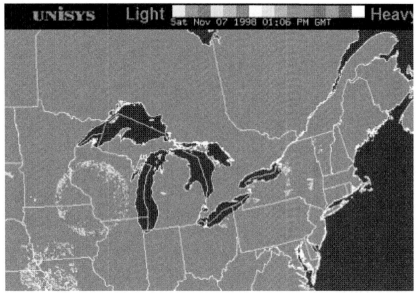

Figure 113 - US Radar Anomaly 07 November 1998

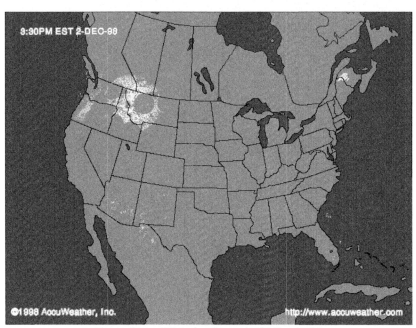

Figure 114 - US Radar Anomaly 02 Dec 1998

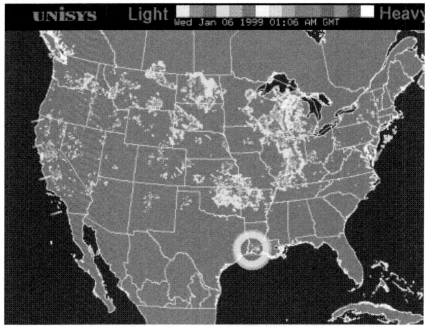

Figure 115 – US Radar Anomaly 06 January 1999

We will look at one suggested explanation at the end of this section.

Switzerland

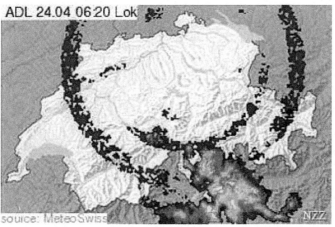

Figure 116 – Swiss Radar Anomaly 24 April 1999

A report posted on Macedonia Online[201] on 05 May 2009 explains:

> On the morning of the 24th of April 2009 employees at MeteoSwiss (Switzerland's Federal Office of Meteorology and Climatology) got the surprise of their lives. Huge circular fields hovering over Switzerland were picked up by their radars. The question was asked as to whether a massive UFO was in the upper-atmosphere above the scenic mountainous nation or perhaps these circles were signs of an imminent hurricane or maybe some sort of electromagnetic radio frequency/microwave technology is in use to cause such an anomaly.

Radar Bloom, Ducting and Anomalous Propagation

An article on accuweather.com[202] discusses the causes of these sorts of anomalies and the author, Jesse Ferrell writes:

> There are a lot of interesting anomalies that you may see on displays that show NEXRAD (or any kind of) weather radar data. Some are caused by software, some are caused by the radar misinterpreting what it sees. None are worth some of the conspiracy theories that non-scientists have come up with.

The article later states the author communicated with someone at the National Oceanic and Atmospheric Administration (NOAA) in the USA.

> Joe Chrisman from the ROC (Radar Operations Center) Engineering Branch, who explained:
>
> "When the sun goes down and the surface begins to cool, the change in refractive index in the lowest few (to several) hundred feet of the atmosphere tend to bend the radar beam toward the surface. This bending holds the radar beam near the surface for extended distances, where it encounters scatterers that would not normally be available above the boundary layer. These scatterers include insects, bats, aerosols, particulate matter, etc., and account for the increased radar return referred to as 'radar bloom.'"
>
> To decode that answer a little, what he's saying is that it is, in fact, superrefraction that causes radar bloom.

This explanation refers to effects seen after the sun goes down. I therefore must point out that any such effects do not completely explain these ring anomalies. The example radar images from the USA show a time stamp – and these times are during both daytime and night time. Also, if you view the animated sequence, each anomaly persists for a certain length of time – 30 minutes to one hour. Closer examination of the radar images reveals several types of anomalies.

Weather Anomalies Captured by Trackers

On close inspection of some of the time lapse videos from the Leeds tracker (which had the longest view to the horizon), some strange cloud movements and formations were seen. Obviously, these are best illustrated by watching the video itself, though I have tried to describe a few stills below.

Leeds - 08 December 2013

On this particular day, the view showed thick cloud covering most of the visible sky area – with clearer weather in the distance (to the north).

Figure 117 – This view persisted for about 90 minutes until precisely 12:00 (noon).

Figure 118 – 12 Noon – cloud starts to clear from the bottom to the top of the image.

In a 30-minute period, the lower cloud layer clears to the left (east) and reveals two higher cloud layers – a lower, thicker dark layer and an upper, lighter coloured layer.

Figure 119 – Higher, persistent, cloud layer is then revealed.

An hour later, the cloud configuration looks remarkably similar, with slight easterly movements of the upper and lower cloud banks.

Figure 120 – A similar view persists for over two hours.

Figure 121 – Same cloud bank remains for over 1 hour.

Why would these cloud banks persist over a fixed point on the ground, when there are no mountain or high hill ranges below them?

In other videos, it seems there are banks of "persistent cloud" formations – cloud forms and reforms over a fixed point on the ground and the "rest of the weather" seems to flow past these clouds. Again, these effects only become obvious from watching the time lapse videos. [203]

Linear Weather Anomalies

Whilst even the C. J. P. Cave book shows linear cloud features in images from 1943, such as those shown below, we will examine some more modern examples here:

Figure 122 – Cirrocumulus waves at sunrise (Figure 12 from the CJP Cave book)

Figure 123 – Altostratus with altocumulus bands at a lower level (Figure 22 from the CJP Cave book)

Most of us have seen linear clouds like this, and banding – we have probably seen cloud patterns resembling the pattern of the sand on the beach when the tide has gone out – or a "mackerel sky[204]" to give it the more usual name.

Figure 124 – Mackerel Sky

However, we will look at some other strange examples here. I am not sure whether linear cloud features have become more common, or whether I am just noticing them more. On several occasions, I have witnessed a kind of "shelf cloud," where there appears to be a thick cloud bank with a straight edge – as pictured below.

Figure 125 - Linear Cloud Bank - 18 June 2009, Long Eaton, UK

Figure 126 - Linear Cloud Bank - 25 December 2007, Borrowash, UK

In a video posted online in 2011 by Matthew Marley, we see this[205]:

Figure 127 - Straight-edged cloud bank (video still) - Matthew Marley - 2011

Quite a few similar examples can be found, but this seems to be one of the clearest. In a satellite image of Montana taken on 07 June 2004 below, we see a square-looking cloud (middle right hand side) ...

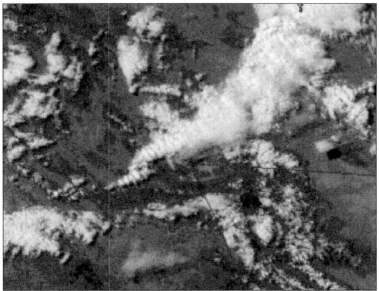

Figure 128 – Satellite image – June 7, 2004–0108Z Southern Montana/East Idaho/NW Montana.

Figure 129 – Enlarged section of above figure.

Scott Stevens a former TV Weatherman from Idaho, in 2004, wrote[206]:

> *June 7, 2004–0108Z Southern Montana/East Idaho/NW Montana. This image was huge for me. Finally convincing me that this project wasn't limited to individual 'events' but was everywhere all the time. Soak up all the oddities, squares, and clouds at 90-degree angles. the whole storm had to have been digitized into individual cells. I've never looked at the sky the same since."*

A further example of geometry in the cloud deck can be seen in this 2009 image below[207]. It is particularly unusual when seen in the context, as it appears to be near the centre of a cyclone:

Figure 130 – Square cut-out in cloud bank – 2009.

UK – Infra Red Photo Weather Anomalies – 10 September 2010, Met Office

Note criss-cross lines – west of Bay of Biscay (bottom left of photo) – original source now archived[208]. This criss-cross pattern appeared between 0400 and 1600 hours.

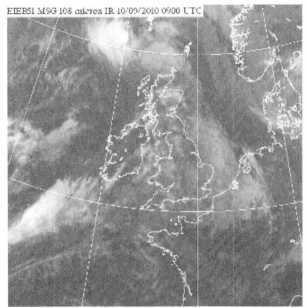

Figure 131 – Unusual trail pattern west of Bay of Biscay, 10 Sep 2010

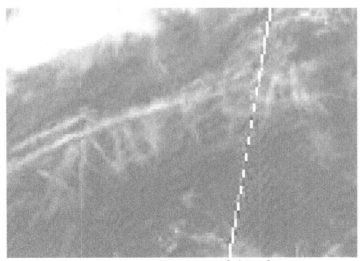

Figure 132 – Enlarged section of above figure.

This can also be seen in a video of frames[209]. How is it that we are seeing this sort of pattern caused by aircraft trails? Is it simply movement of air masses following the passage of civilian aircraft? I am having a hard time accepting this as the full explanation for this phenomenon.

Some people will just say "nature can do anything …."

Hole Punch or Fallstreak Hole Clouds

Since about 2003, a few pictures of these unusual clouds have been posted online and I find them quite extraordinary. Here are a couple of examples:

Figure 133 – This image was taken the morning of November 17, 2004 from a plane departing New York's JFK airport. [210]

Figure 134 – 2004 'Hole in sky' amazes scientists – Alabama[211]

An Explanation?
There are more interesting pictures in an article titled "Hole-punch clouds are made by jets" on "Earth and Sky", where we read[212]

> *Andrew Heymsfield of the National Center for Atmospheric Research spoke with EarthSky some years ago, when his study first appeared. He told us:*
>
> *"This whole idea of jet aircraft making these features has to do with cooling of air over the wings that generates ice."*
>
> *His team found that – at lower altitudes – jets can punch holes in clouds and make small amounts of rain and snow. As a plane flies through mid-level clouds, it forces air to expand rapidly and cool. Water droplets in the cloud freeze to ice and then turn to snow as they fall. The gap expands to create spectacular holes in the clouds. He said:*
>
> *"We found an exemplary case of hole-punch clouds over Texas. From satellite imagery, you could see holes just pocketing the sky, holes and long channels where aircraft had been flying at that level of the cloud for a while."*

The referenced author Andrew Heymsfield[213] has 406 papers listed on the NCAR & UCAR website [214]. (He is listed as a "Senior Scientist" though no details of his qualifications are given.)

> *The University Corporation for Atmospheric Research is a non-profit consortium of more than 100 North American member colleges and universities focused on research and training in the atmospheric and related Earth system sciences.*

His paper titled "Aircraft-Induced Hole Punch and Canal Clouds – Inadvertent Cloud Seeding"[215] is very interesting. It is of a much higher standard than the Herndon Paper or the West et al "expert opinion" paper, and contains details of specific observations and detailed analysis of those observations. The paper opens with these words:

> *Passage of commercial aircraft through supercooled altocumulus can induce freezing of droplets by homogenous nucleation and induce holes and channels, increasing the*

> *previously accepted range of temperatures for aviation induced cloud effects*

The first page also contains this interesting satellite image:

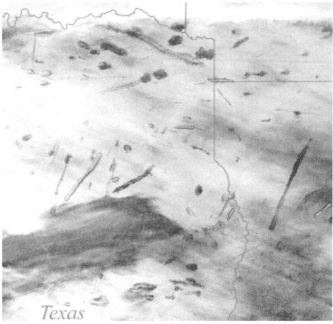

Figure 135 - from the Heymsfield Paper - Image of hole punch and canal clouds captured by the Moderate Resolution Imaging Spectroradiometer (MODIS) on NASA's Terra satellite at 1705 UTC 29 Jan 2007 (image courtesy of NASA Earth Observatory and specifically J. Schmaltz). State boundaries are shown.

The paper doesn't comment on what looks to me like a rectangular area of thinner cloud towards the lower left of the image (just above the word "Texas"). We can also observe faint cloud lines running parallel with the shorter side of this partial rectangle.

On page 12 (or 764), the paper includes some detailed measurements and even photographs showing the apparent evolution of one of these cloud features:

Weather Anomalies

Figure 136 – from the Heymsfield Paper – Sequence showing the production of what appears to be a contrail in or below an altocumulus deck, the formation of ice and a canal cloud, and the dissipation of the altocumulus layer with the residual ice layer.

Whilst we do see a strange-looking wispy cloud – which could perhaps be described as halfway between a cirrus cloud and a contrail, it isn't a "hole punch cloud."

A page on "weather.gov" shows a photo by Samantha Weise, with the caption as written:

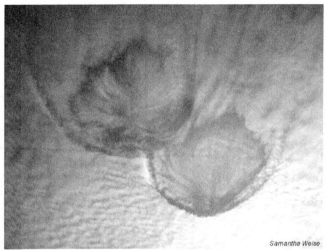

Samantha Weise

Figure 137 - Source caption/explanation: "*Hole punch cloud - The source of the ice crystals was exhaust from jets. Visible satellite imagery on November 15 showed aircraft dissipation trails (distrails) above central and south-central Wisconsin. Particles in the aircraft exhaust acted as ice nuclei, causing the ice crystals to grow and then fall, creating the punch holes and streaks in the main cloud layer.*"

Is this the correct explanation? Why don't we see persistent contrails bisecting or touching all these features? (Yes, we can see evidence of them on some photos.)

Other Examples

These seem rather different in character:

Figure 138 - Battle Creek Michigan 21 September - 2004 nearing sunset[216]

Weather Anomalies

Figure 139 -2004 - Orlando Florida late in the day prior to Christmas Eve. Blooming "contrails" evident with a striking oval imprint deposited into the cloud field. [217]

Though we can see in both these examples contrails/chemtrails are nearby or even crossing over the anomaly, we cannot determine their true relationship from photographs alone.

The details of these next two images are not known, though from the file names they were given, they were probably taken on 12 December 2005.[218] No contrails in sight:

Figure 140 - Circular clouds - USA - Dec 2005

Figure 141 – Circular clouds – USA – Dec 2005

Many more interesting images can be found online – including on a page by Glenn Boyle called "Strange Days Strange Skies."[219]

From what I have read, none of the explanations for these anomalies includes a reference to climate changes caused by human industrial activity.

Asperatus Clouds

This term appears to have been coined in about 2008 or 2009 to cover a particular type of cloud that has been observed in several places around the world.

Figure 142 – "Asperatus" Clouds, UK, 02 June 2009

An article posted on the UK *Daily Mail* website on 02 June 2009[220] includes the picture above, by Ken Prior, along with a number of others from around the world. A similar scene to the one above was witnessed by "Kristi" on 25 March 2009 in Fort Worth, Texas.

Figure 143 -"Asperatus" Clouds, "Kristi" on 25 March 2009 in Fort Worth, Texas.

These clouds were also recorded on video by another witness.[221]

Persistent Linear Rainbow Effects and Repeated Sundogs

Whilst at home on 15 June 2009 I took the photograph below. The strange "angled rainbow coloured lines" persisted for about 15 minutes.[222]

Figure 144 -15 Jun 2009 – Borrowash, UK

Personally, I have noticed many more of these strange-looking cloud formations over the last few years.

Figure 145 – Sundog – Borrowash – 11:09am, 11th November 2004

Many people mistake these Sundogs for something else (such as a UFO or "Planet X"). Though the phenomenon is well understood[223], I have observed them more often since the 1990s (but maybe I just notice them more).

Conclusion

What I have shown you in this chapter are many different types of weather anomaly. They have been seen

- on ground-based photos
- on weather RADAR
- on satellite images

A number of these images seem to show unusual geometry, and though in some cases we can attribute this to natural processes (such as with the Mackerel Sky), I contend that something else is going on. An argument can be made that because more and more people have their own video and still cameras, they are recording these more regularly. However, weather RADAR has been in use for over 20 years, and satellite imagery for much longer, but few scientists offer real explanations for these observations and they are usually just ignored.

Collectively, I think these observations show that the earth's atmosphere has been changing. This change is either part of a natural process or it has been artificially induced, or a combination of these. Due to the geometry associated with some of the effects witnessed, it is extremely unlikely that the sole explanation for these phenomena is simply that there have been changes in the concentration of the component gases of the atmosphere from industrial or similar processes!

In the following chapters, I will go through some of the evidence which suggests that large scale, sophisticated but covert geoengineering has been in use since at least 2001.

CHAPTER ELEVEN
GEOENGINEERING – CIVILIAN AND MILITARY

'Tis strange – but true; for truth is always strange;
Stranger than fiction: if it could be told,
How much would novels gain by the exchange!

Lord Byron (1788–1824) *Don Juan*, XIV

Space for Notes Below

Don't Talk About Things in the Past!
In Chapter Four, we discussed legislation and government studies which relate to geoengineering – or weather modification, to use its other name. There is therefore an implicit assumption that geoengineering is not currently in use. Its previous use has not been discussed in the documentation pertaining to the mitigation of non-existent Anthropogenic Global Warming, as far as I can tell.

However, if one studies available records, documentation and literature, it appears that a number of attempts at geoengineering have been made in the last 100 years. This chapter is not meant to be a comprehensive review of the available historical literature.

Charles Hatfield
In 1915 the San Diego city council approached Charles Hatfield to produce rain to fill the Morena Dam reservoir – he claimed to be a rain maker. Hatfield, with his brother, built a 20-foot (6-metre) tower beside Lake Morena and was ready early in the New Year – they used a secret mixture of chemicals. Starting on 5th January 1916, it rained for 15 days and caused floods and dams to burst! Hatfield was blamed for the damage[224]. Hatfield's story inspired the 1956 Burt Lancaster film "The Rainmaker"[225].

Figure 146 – Charles Hatfield at work.

Hatfield's work was even discussed in a JSTOR article in 2015, during a worsening of drought conditions in California, USA[226]:

> *As California endures its worst drought in 1,200 years, residents of the Golden State are turning to extreme—and desperate—measures to quench their collective thirst. Sun-*

> baked farmers are hiring "water witches" to divine underground water sources with forked branches, while a company called Rain on Request has pledged to end the drought by building electrical towers that would induce rainfall by ionizing the atmosphere. When California found itself in a similar parched position exactly 100 years ago, the city of San Diego did something that seems even more bizarre—it hired a rainmaker. The thing is, it might have worked. After Charles Mallory Hatfield began his work to wring water from the skies, San Diego experienced its wettest period in recorded history.

A 25 May 2015 story in the Los Angeles Times repeats many of these details.[227]

Cloud Seeding in the 20th Century

In relation to the "spraying of material out of aircraft", there have been quite a few occasions where this has, indeed, taken place. Research seems to have started in the 1940s and the first cloud seeding experiments were done in the 1950s. It seems the two most common materials used for cloud seeding are dry ice (ironically, this is frozen carbon dioxide) and silver iodide.

One scientist associated with the invention of cloud seeding is Irving Langmuir – who was the inventor of the high-vacuum tube.[228]

Figure 147 - Irving Langmuir

He was involved in conducting the first attempt at cloud seeding in Massachusetts in 1946.[229] A plane "seeded" a cloud with crushed dry ice and snow began falling out of that cloud. This was done as a result of an accidental discovery by Langmuir's colleague, Vincent Schaeffer, following his usage of an experimental "cold box"[230] where droplets of super-cooled liquid formed (below freezing point but not solid) when Schaeffer breathed into it. Langmuir recruited Bernard Vonnegut to help with the research his team was doing.

Figure 148 - Bernard Vonnegut in 1988

Atmospheric scientist Bernard Vonnegut is credited with discovering the potential of silver iodide for use in cloud seeding. This property is related to a good match in lattice constant between the two types of crystal.[231]

It is important to mention that silver iodide is mostly used for hail suppression, while a newer technique called hygroscopic seeding[232] is used for the enhancement of rainfall in warm clouds. These techniques neither create nor attract or move around moisture – they merely encourage precipitation when it is present in the atmosphere.

Hail storms can cause a lot of damage to crops and property in certain parts of the USA. Perhaps for insurance-related reasons, a US company called Weather Modification Inc will perform a cloud-seeding service. [233] The company's website states: [234]

> **Feasibility Studies** – Weather Modification, Inc., has been involved in multiple programs with the goal of identifying the potential benefits derived from the application of cloud seeding techniques. Our feasibility studies focus on the cloud structures and patterns in the project area. Gathered information assists in identifying the cloud seeding technology that's best suited for the project.

Disclosed Military Projects: Cloud Seeding Projects

It was only in June 2013 that I became aware that in John F Kennedy's Address at the U.N. General Assembly on 25 September 1961 [235], he said:

> *"We shall propose further cooperative efforts between all nations in weather prediction and eventually in weather control."*

Can we ask, in 2017 – almost 56 years on from his statement – has weather control been achieved ...?

Here, I briefly cover two disclosed geoengineering projects from the 1960s which show that such activities were undertaken, for various reasons. These projects were named "Stormfury" and "Popeye."

Project STORMFURY – Ben Livingston

"Stormfury" was an experimental project to modify the action of hurricanes and reduce the damage they caused.[236]

Experiments were carried out between 1962 and 1983 and involved Ben Livingston[237]. Livingston was interviewed in 2011 by Alex Jones. (Jones runs a site called Infowars.com but has proved to be untrustworthy in a number of areas – I have written about this in my free e-book "911 Finding the Truth). Livingston maintains that their efforts to reduce the damage caused by hurricanes were successful - hurricane wind speeds were reduced by cloud seeding. Others suggest that the seeding had no real effect[238].

Figure 149 - Ben Livingston – military cloud-seeding pilot.

Livingston also stated he was involved in Operation Popeye, which was a secret US Military Operation to create rainfall over the Ho Chi Minh Trail in Vietnam[239]. Some details of this are revealed on page 112 of a declassified document of the minutes of "Weather Modification Hearings" which took place on January 25 and March 20, 1974. An article on *Opsec News* provides a very useful overview of the operation[240] – the US Military kept their operation secret for a number of years, but a leaked memo triggered the secret hearings.

"Eco-Terrorism"?

It is actually, then, not much of a secret that the US military like to be able to change or control the weather. When you start to explore this area, you will quickly find this quote by Secretary of Defence, William Cohen, who talked of "eco-terrorism" in 1997.[241]

Figure 150 - William Cohen, former US Defence Secretary.

"Others [terrorists] are engaging even in an eco-type of terrorism whereby they can alter the climate, set off earthquakes, volcanoes remotely through the use of electromagnetic waves ... So there are plenty of ingenious minds out there that are at work finding ways in which they can wreak terror upon other nations ...It's real, and that's the reason why we have to intensify our [counterterrorism] efforts."

It is interesting to see that Mick West's "Metabunk" site claims to have "debunked" the fact that Cohen said this[242] – when the transcript is quite clear on the archive.defense.gov website. However, West's site – intentionally or unintentionally, forms part of the perception management framework.

Of course, the Cohen statement came four years before the events of 9/11 – which we will cover later. Having researched 9/11 intensively, I have concluded that Cohen's statement is another example of mischaracterising a threat – i.e. the real terrorists are the global control groups, not "rogue nations."

"Weather as a Force Multiplier: Owning the Weather in 2025" – August 1996

A document with this title is hosted on the USAF Website[243] (www.af.mil) It states that it is:

> *A Research Paper – Presented to Air Force 2025 by Col Tamzy J. House, Lt Col James B. Near, Jr., LTC William B. Shields (USA), Maj Ronald J. Celentano, Maj David M. Husband, Maj Ann E. Mercer, Maj James E. Pugh.*

It has a disclaimer on page 2, which reads as follows:

> *2025 is a study designed to comply with a directive from the chief of staff of the Air Force to examine the concepts, capabilities, and technologies the United States will require to remain the dominant air and space force in the future. Presented on 17 June 1996, this report was produced in the Department of Defense school environment of academic freedom and in the interest of advancing concepts related to national defense. The views expressed in this report are those of the authors and do not reflect the official policy or position of the United States Air Force, Department of Defense, or the United States government.*
>
> *This report contains fictional representations of future situations/scenarios. Any similarities to real people or events, other than those specifically cited, are unintentional and are for purposes of illustration only.*
>
> *This publication has been reviewed by security and policy review authorities, is unclassified, and is cleared for public release.*

So, can this document be at least taken as a "statement of intent"? Or, should we add "two and two together" and consider that black programs possess certain technologies that are years in advance of their white world counterparts? Perhaps you can go back to page 10 above and reread Eisenhower's statement.

The HAARP Project and Ionospheric Heaters

If you are having a conversation with someone about their disbelief in officially sanctioned propaganda about global warming and climate change and they say the word "chemtrail", it is likely that the word (acronym) "HAARP" will only be a few breaths away ... This is because it has now become an almost knee-jerk reaction for them to say that "HAARP is affecting global weather." Whilst some information suggests that this might be partly true, it is by no means certain.

Figure 151 – HAARP Array in Gakona, Alaska.

HAARP is the High Frequency Active Auroral Research Program[244]. It is described as

> ... the world's most capable high-power, high frequency (HF) transmitter for study of the ionosphere. The principal instrument is the Ionospheric Research Instrument, a phased array of 180 HF crossed-dipole antennas spread across 33 acres and capable of radiating 3.6 megawatts into the upper atmosphere and ionosphere. Transmit frequencies are selectable in the range of 2.7 to 10 MHz, and since the antennas form a sophisticated phased array, the transmitted beam can take many shapes, can be scanned over a wide angular range and multiple beams can be formed. The facility uses 30 transmitter shelters, each with six pairs of 10 kilowatt transmitters, to achieve the 3.6 MW transmit power.

3.6 MW is a lot of energy – enough to power a small town, although there is indeed a vast amount of energy in the atmosphere (as already implied on page 49). The HAARP FAQ page is worth reading in full. It continues:

> **Why Was HAARP Developed?**
>
> Between 1990 and 2014, HAARP was a jointly managed program of the United States Air Force (USAF) and United States Navy. Its goal was to research the physical and electrical properties of the Earth's ionosphere, which can affect our military and civilian communication and navigation systems.

A posting titled "HAARP again open for business" dated 03 September 2015 on the HAARP website states[245]:

> In mid-August, U.S. Air Force General Tom Masiello shook hands with UAF's Brian Rogers and Bob McCoy, transferring the powerful upper-atmosphere research facility from the military to the university.

In a book *Angels Don't Play This HAARP*,[246] Jeanne Manning and Nick Begich discuss other potential uses of HAARP, such as Earth Penetrating Tomography. The same method is mentioned in a 2002 research proposal overview by John Pike titled "HAARP – Detection and Imaging of Underground Structures Using ELF/VLF Radio Waves," which on the FAS website states:[247]

> Several distinct methods for ELF/VLF generation are available to support these efforts. Proposers are encouraged to consider including the controlled ELF/VLF sources provided by the 960KW HF transmitter of the HAARP, presently under construction outside Gakona, Alaska and the HIPAS facility located near Fairbanks, Alaska. PL/GPS is the program manager for the HAARP facility. The Office of Naval Research controls the HIPAS facility. Both sites will be available to support the research efforts under this PRDA.

This idea that HAARP could be involved in "earth penetrating" activities might be what has led some to claim there is a link between the October 2009 Haiti quake and HAARP activity[248].

Bernard Eastlund's patent[249] on which HAARP is based should be studied, bearing in mind several similar but smaller facilities *are already in operation*. (For example, HAARP in Alaska and EISCAT in Norway. [250]) Other facilities are said to exist in Puerto Rico – at Arecibo [251] and there was a proposal for an ionospheric heater in Newcastle, Australia[252].

HAARP's own FAQ, mentioned above, acknowledges that the action and effects of its operation are normally invisible. However, we can find an article in "Nature" from 02 October 2009 titled "Artificial ionosphere creates bulls eye in the sky" : [253]

> But in February last year, HAARP managed to induce a strange bullseye pattern in the night sky. Instead of the expected fuzzy, doughnut-shaped blob, surprising irregular luminescent bands radiated out from the centre of the bullseye, says Todd Pedersen, a research physicist at the US Air Force Research Laboratory in Massachusetts, who leads the team that ran the experiment at HAARP.
>
> The team modelled how the energy sent skywards from the HAARP antenna array would trigger these odd shapes. They determined that the areas of the bullseye with strange light patterns were in regions of denser, partially ionized gas in the atmosphere, as measured by ground-based high-frequency radar used to track the ionosphere.
>
> The scientists believe that these dense patches of plasma could be gas that was ionized by the HAARP emissions. "This is the really exciting part — we've made a little artificial piece of ionosphere," Pedersen says.

Is it really just an experiment by over-enthusiastic scientists to make interesting glows in the sky ...?

An Earlier HAARP Facility at Poker Flats, Alaska

According to Gary Vey's statements in a Red Ice Radio interview in January 2011, [254] he was contacted by two whistleblowers regarding this "other" HAARP facility. He explained that "the HAARP that everyone knows about is the one at Gakona, Alaska." The whistleblowers told him there was a second secret HAARP facility at Poker Flats – which had harmed some local people during its operation. Poker Flats is officially a NASA rocket launch facility[255]. There is a disclosed connection between the two sites in a presentation by Dr Bob McCoy, Director, Geophysical Institute, University of Alaska Fairbanks. (See page 17 of his presentation "Space Weather Effects with the Flip of a Switch.") [256]

The Poker Flats HAARP facility is also mentioned in a posting on Gary Vey's website ominously titled "Weapons of Total Destruction."[257]

Summary

In this chapter, I have briefly shown past examples of disclosed geoengineering activities. In one case, (Project Popeye) the activity was deliberately kept secret. The information about the HAARP project also seems sketchy in places and whilst many have made wild claims about it, I am not confident that HAARP is being used for benign purposes.

There is plenty of additional research on the use of geoengineering for military reasons – such as that documented in Dr Rosalee Bertell's book *Planet Earth: The Latest Weapon of War.* [258]

CHAPTER TWELVE
WILHELM REICH, CLOUD BUSTING AND ORGONE ENERGY

"Man's right to know, to learn, to inquire, to make bona fide errors, to investigate human emotions must, by all means, be safe, if the word 'freedom' should ever be more than an empty political slogan."

Wilhelm Reich

Space for Notes Below

Hang on to Your Hats!

In the preceding chapters, we have shown that changes in the weather can be affected or induced by chemical (cloud seeding) or energetic (HAARP) means. Is it possible that someone else stumbled across a way to affect the weather with simple equipment?

In this chapter, we will consider the work of Wilhelm Reich, though I urge you to read Reich's books,[259] some of which are now available in electronic form and are therefore more cheaply accessible.

Figure 152 – Wilhelm Reich

Finding out about Reich

Wilhelm Reich is someone I consider to be in the same pantheon of brilliant scientists as Nikola Tesla, but Reich's name is not nearly as well-known. I first became aware of Reich's work in about 2003 or 2004, when I started to investigate a range of alternative knowledge topics.

We have already discussed what some people call "chemtrails" – persistent jet trails in the sky. We have considered the idea that these trails are toxic spray. At the moment, I am not sure what they are, [260] but it was in researching persistent jet trail/chemtrail phenomena that I first heard about Wilhelm Reich. This is because Reich had developed and worked with what he called "Cloudbuster" devices in the 1940s and 1950s.

I later discovered that British vocalist Kate Bush's song *Cloudbusting* [261] was inspired by Reich's work [262]. It was based on [263] the story related in the book written by Reich's son – *A Book of Dreams*. [264] (I was later given to wonder if Kate Bush's interest in Reich extended beyond *A Book of Dreams*, because in a 1979 interview she was asked a question about an alternative career and she answered that she was interested in being a psychiatrist – and Wilhelm Reich started his research in Psychiatry. [265]

Figure 153 – Peter Reich's "A Book of Dreams" – 2015 edition.

Peter Reich's book was originally published in 1974, and reprinted in 1989. Renewed interest in this book meant that an ebook edition and new paperback edition were published in March 2015. Although this book is more about Peter Reich's life and how it was obviously strongly influenced by his father, there are certain important quotes which I found to be quite illuminating, such as when Peter Reich discusses why there are so many attacks on Wilhelm Reich and his work.

> "It is an emotional plague that comes from within. It kills people emotionally and makes them keep their belly tight. It makes them lie and slander and spy. This emotional plague is more vicious than the black plague because the people do not want to be cured. They strike out in rage at one who tries to cure it because they have been sick so long they think the sickness is health. And that is why they are attacking me. I am trying to tell them that they don't have to hate and they hate me for it."

To those of us that have tried to pass on to others some of the challenging facts we have learned, this description should have us nodding in agreement. Reich had come from a background of studying people's behaviour and had realised some of the fundamental problems which arise when their entire world view is challenged – even when such a challenge could lead to great improvements in their lives. The "emotional plague" which Reich identified seems to have gotten worse since his death – and I would argue that this is not a natural or entirely accidental process. The way that society has been and is being "configured" allows the "emotional plague" to thrive. (For more information on this, see chapter 4 (and elsewhere) of Reich's book "Ether, God & Devil & Cosmic Superimposition."

Observing Nature

In his book *Ether, God and Devil* Reich makes revelations which, in my opinion, go quite a long way to explaining the way the world is now. As I understand his conclusions, he says that people have become detached from observing the natural world and as this happened, they lost the ability to think for themselves. Gradually, their own thoughts and understanding of the world were eroded by religious beliefs and practices and so observations of the real world became secondary to what people were told about "god" and how they should live their lives. People were prevented from understanding how the life force energy flows through their being and how it interacts with all living things and creates real effects. As we started to live in larger and larger settlements and fewer and fewer of us worked on and in the land, we stopped observing natural processes and our consciousness changed and became less experiential.

Reich regarded religious fundamentalism – and what some have come to call "scientism" – with similar levels of scepticism. In his chapter "Animism, Mysticism, And Mechanistics" he describes the limitations of mechanistic science, writing:

> *The history of science leaves no doubt that the living process was not allowed to be studied; that through thousands of years it was the mechanistic-mystical structure of the human animal that excluded from all research, by absolutely all conceivable means, the cosmic foundations of the living process.*

Reich's Observations

Reich followed a path of ongoing observation and experiment and he was able to show that orgone energy is real – it is everywhere, not just part of the processes which take place in human sexual activity. I have to wonder if one of the reasons why Reich's work and research is less well known and understood is because much of it is so fundamentally challenging. It explores the areas that "make us like we are." What is probably more well-known about Reich is that he worked with Sigmund Freud – as a clinical assistant. It was Reich's ongoing clinical work that led him to investigate "sexual energy." Freud had also researched this same energy and, apparently, he made similar findings[266] …

> Freud had discovered that neuroses are caused by the conflict between natural sexual instincts and the social denial and frustration of those instincts. Freud had also hypothesized the existence of a biological sexual energy in the body. He called it "libido," and described it as "something which is capable of increase, decrease, displacement and discharge, and which extends itself over the memory traces of an idea like an electric charge over the surface of the body."
>
> But as the years passed, Freud and his followers diluted much of this concept, reducing the libido to little more than a psychological energy or idea. By 1925, Freud had concluded that "the libido theory may therefore for the present be pursued only by the path of speculation."

Perhaps members of the "Old Boys Network" whispered in Freud's ear …? By this, I mean that Reich was onto something *big* … really big. Freud, like most other scientists, ran away (or was he *told* to run away?) from these discoveries and areas of research. Reich, unlike most other scientists knew that what he was discovering was important, so he pursued it – with passion, reason and intensity.

If this sounds far-fetched, consider what happened to Dr Martin Fleischmann, Dr Stanley Pons and others when they discovered a new energetic process in electrochemistry. Fleischmann experienced what he called "pathological criticism" in the field of what became known as "Cold Fusion" research. This happened (and continues to happen) when scientists denied there were any unusual effects being witnessed and documented – even when these effects had been recorded tens or hundreds of times – by different scientists in different countries.

The Taboo of Reich
Even when I had found out about Reich's research into weather modification and then also became aware of the connection to Freud, I still had no awareness that the College of Orgonomy [267] or the Wilhelm Reich Infant Trust still existed [268]. From the little that I had read, I assumed that Reich's work had been successfully suppressed, to the point that no one was actively promoting, let alone using or benefitting from, any of his research. Perhaps I just hadn't read enough, or wasn't paying close enough attention, but after reading a description such as the one below [269], I think I could be forgiven for assuming little or nothing had survived after the 1950s:

> On March 8, 1957, four days before he was taken to a federal prison, Wilhelm Reich signed his Last Will and Testament. By this time his orgone energy accumulators and many of his publications had already been banned and destroyed by order of a United States Federal Court injunction, starting on June 5, 1956 when three orgone energy accumulators were destroyed outside of Reich's Student Laboratory at Orgonon in Rangeley, Maine.
>
> Three weeks later, several boxes of his publications were burned under the supervision of Food and Drug Administration agents outside the Student Laboratory. A month after that, in July, the panels for about fifty orgone accumulators were dismantled in the town of Rangeley, Maine by the local contractors who had built them.
>
> And exactly one month after that–on August 23, 1956–several tons of Reich's publications, including the titles of 10 hardcover books as well as medical and scientific research bulletins and journals, were burned under FDA supervision at a New York City municipal garbage incinerator on Gansevoort Street.

Orgone Energy
Reich performed simple experiments to find out more about the "life force" energy which he had seemingly discovered during his treatment of patients. In the 1940s, he built what he called an "Orgone Accumulator." He realized that a simple arrangement of layers of organic and inorganic materials could "trap" this Orgone energy.

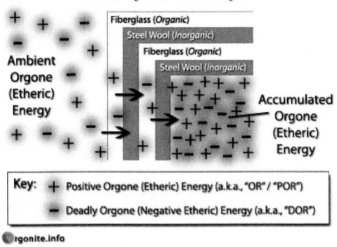

Figure 154 – Orgone Accumulator – operation.

In one experiment with an Orgone Accumulator (ORAC), described in Part IV ("The Objective Demonstration of Orgone Radiation") of Reich's book *The Cancer Biopathy*, Reich measured an anomalous temperature difference. Dr James De Meo and others have confirmed these findings. [270]

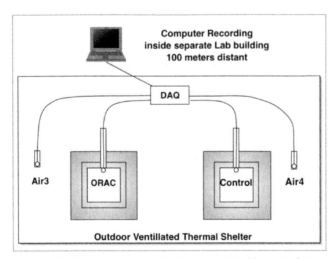

Figure 155 – De Meo's ORAC reproduction: General layout of the To–T Experiment. Thermistors measure the temperature in the upper interior of both ORAC and Control enclosures, with separate air temperature measurements.

De Meo writes in the "Abstract" of his paper:

> Experimental investigations were undertaken by the author, of the thermal anomaly (To-T) inside the orgone energy accumulator (ORAC), a phenomenon firstly observed by the late Dr. Wilhelm Reich, who invented the ORAC device. This thermal anomaly, by the theory of Reich, is produced from the rarefied motional-pulsating orgone energy continuum which is concentrated Inside the ORAC, producing a frictional thermal heating of the air ....
>
> ... The experiments confirmed Reich's claims of a slight spontaneous heating effect inside the ORAC, which has no known energy source by classical "empty space" determinations.

Reich also observed the visual and radiative effects of this subtle energy. So, Reich confirmed there was another type of subtle energy that could be detected. Could it be directed or manipulated?

Cloudbusting

In Jerome Eden's book *Planet in Trouble: The UFO Assault on Earth?* [271]; First edition (June 1973) we can read:

> Underlying the Cloudbuster is Reich's discovery of the Orgonomic potential – the flow of Orgone energy from the weaker to the stronger OR system. Water in a deep well, flowing brook, lake or ocean, contains a higher OR potential than does the atmosphere or passing clouds. The higher concentration of OR, following the orgonomic potential, will therefore attract the OR of a weaker system to itself. Energy is thus "drawn" from the weaker to the stronger system. Knowing this fundamental law, an orgone engineer can either lower or raise the OR potential in the atmosphere by judicious drawing operations.

Eden also writes:

> 1940, Reich observed that hollow metal pipes appeared to influence wave motion on the surface of a Rangeley lake when the pipes were directed at the water. This phenomenon seemed incomprehensible, and Reich all but forgot about it

until years later when the DOR emergency at his Orgonon laboratory pressed him to seek relief. As the DOR emergency became unbearable, Reich pointed several metal pipes, about 10 feet long and 2 inches in diameter, at a concentration of DOR in the skies above his laboratory. The pipes were grounded via hollow "armor" cable, known in the electrical trade as "BX" cable, to a deep well in the property.

Observers noted an incredible result of this procedure. The dirty, nauseating DOR clouds began to break up and diminish in size! When the pipes were pointed in a direction opposite to the normal west-to-east movement of the atmospheric OR energy flow, a breeze would start up, as if out of nowhere. From this early, simple and effective beginning grew the science of Cloudbusting and CORE (Cosmic Orgone Engineering), that branch of Orgonomy which deals specifically with man-made weather phenomena, including the destruction and formation of various types of clouds and cloud cover and all induced weather phenomena, such as climatic and atmospheric changes encompassing humidity, barometric pressures, production and cessation of rainfall or snowfall, the "turning" of tornadoes and hurricanes away from paths of destruction, and the greening of deserts and parched areas.

Figure 156 – Reich operating a Cloudbuster

In Reich's final book *Contact with Space* (which was not properly finished), on pages 90-91 he reports how he observed a connection with Deadly

Orgone (DOR) energy and vapour trails from military jets. He stated that "DOR wiped out trails." He said there were gaps in trails due to the presence of DOR. He even thought the US Air Force were using the trails themselves to check the presence of DOR.

Some people have taken these ideas and developed them into a "Chemtrail Buster", which, it is said, can harness and direct orgone energy to dissipate the chemtrails. [272]

Reproducing Reich

Several other researchers have reproduced Reich's experiments with Cloudbusters, although we should also consider further words from Jerome Eden's book, as follows:

> As with his invention of the Orgone Energy Accumulator, the Cloudbuster is "simplicity itself." Its simplicity and functional operations have eluded scientific minds for centuries. It is, however, just this very simplicity, the ease of construction and use, which makes the Cloudbuster such a potentially dangerous and extremely powerful instrument. No one, under any circumstance, should PLAY AROUND WITH A CLOUDBUSTER! Not only is it extremely dangerous locally, but also it can trigger atmospheric effects of the most powerful kind, including twisters and hurricanes. The extremely serious threat facing humanity, however, must have forced Reich to "tell where the fire-fighting apparatus is"!

This characterization may be somewhat extreme – as quite a few people have now built and experimented with cloudbusters. Although Reich discusses negative effects in his *Contact With Space* book, I have not heard similar reports from other users. Nevertheless, those that have used Cloudbusters extensively do advise caution.

Dr James De Meo

As well as confirming the results of Reich's simple "anomalous temperature difference" experiment, Dr James De Meo has done work with Cloudbusters.

De Meo states: [273]

> "While the cloudbuster makes little sense from the standpoint of classical meteorology, in fact both Reich and several other

> *scientists had published fairly convincing reports of their field experiments since the 1950s. I read those reports, and found them nothing less than amazing. Over subsequent years, I met and trained with two of the scientists who had closely followed Reich's approach, Mr. Robert Morris and Dr. Richard Blasband, and undertook a variety of laboratory experiments to become familiar with his overall discovery. I observed the experiments of others, using large cloudbuster devices, and could hardly believe my eyes at the way the atmosphere responded. When I was a graduate student at the University of Kansas, I finally was able to make my first cautious steps and systematic research into the question."*

Reich observed a working cloudbuster could also have negative biological effects upon the operator, and De Meo has confirmed this personally. This is perhaps what led him to develop cloudbusters that can be remotely operated.

> *"Reich noted certain symptoms of 'overcharge' and other problems which were at the very least a nuisance to this work. At worst, they could be life-threatening. Several persons using cloudbusters have been made severely ill to the point they could no longer do it. I take all kinds of precautions, including use of electrical motors and a remote control so I can stand off by 100 feet and not get close to the device when it is working. We don't know what the long-term health effects are of this device upon the operator, and it might be similar to what the early experimenters in atomic energy suffered through."*

De Meo says he was sickened temporarily on a few occasions, but feels confident about his protective measures. De Meo has undertaken drought abatement work in Namibia, Eritrea, Israel and elsewhere and his results have been impressive, to say the least.[273]

Trevor James Constable

Trevor James Constable (1925-2016) repeated and developed some of the Cloudbuster work of Wilhelm Reich.

> *Trevor James Constable, originally from New Zealand, worked in the Merchant Marines for over 31 years as a Merchant Marine radio officer, he traveled on enormous ships crossing the north eastern Pacific Ocean over 300 times, this gave him*

> the opportunity to experiment with his research based on Wilhelm Reich's work. [274]

Figure 157 – Trevor James Constable

Constable's method differed from Reich's in that he attached his equipment either to a ship or to an aircraft. The ship or aircraft can then be steered in accordance with the flow of orgone energy in the local environment.

Figure 158 – Trevor James Constable and rain making tube.

The most obvious reaction is described by Constable himself: "How is it possible that a simple tube attached to an aircraft could affect the weather so dramatically?"

Like Dr James De Meo, Constable repeated his results in working with "etheric flow" in different countries. A video of his work[275] shows how drought was alleviated in Los Angeles in 1986 and in Singapore in 1988. I recommend you watch other videos about Constable's work[276] in Hawaii and Indonesia and consider his results carefully.

On a website now devoted to his research[277], we can read:

> An environmentally pure technology has been developed by practical men, to deal with drought effectively. No chemicals or radiation in any form are employed. Etheric rain engineering accesses and technologically employs the chemical ether, which permeates and controls the atmosphere via the ether's mighty, yet gossamer-subtle flows.

He continues:

> New-design geometric translators have finally permitted AIRBORNE operations – a long-awaited breakthrough. Drought reversal over large territories became feasible. Airborne tests in Hawaii, and in Malaysia, reveal a stunning potential. Chemicals or electric power are not required. The techniques are simple, environmentally pure – and effective.

Harry Rhodes and His Translators

In August 2017, I met Harry Rhodes – who has built his own version of Constable's Translators. Rhodes has recently claimed success in alleviating years long drought in California, though I have been unable to verify this yet.

Figure 159 – Harry Rhodes and a "Translator" – 05 Aug 2017.

CHAPTER THIRTEEN
"DON'T TALK ABOUT THIS WEATHER ..."

"Errors, like Straws, upon the surface flow;
He who would search for Pearls must dive below."

John Dryden (1631–1700)
All for Love, Prologue

Space for Notes Below

A Storm to Blow Your Mind?

The evidence presented in this chapter seems to be subject to the strongest taboo of all. Almost no one will acknowledge it. Even though it is clear and unequivocal, it is almost never discussed. Neither mainstream nor "conspiracy type" websites will accurately reference this evidence and the book which brought this evidence to light is rarely discussed in any detail – even though the book has been available since late 2010/early 2011.

The weather data I will discuss here was first reported in 2008 by Dr Judy Wood and I contend it is so world-changing and so profound that few people can take in this data – and face the conclusions that should be made... Here is one of the pictures that she first studied in early 2008:

Figure 160 – Hurricane Erin, 11 September 2001

The enlarged section shows the plume of material rising from the destroyed World Trade Centre (WTC) complex. We will discuss this in more detail later in this chapter (or, you can simply trip on over to Mick West's comprehensive "Metabunk" Website and you will see it's all just a coincidence and there's nothing more to analyse in detail ...)

First, Some 9/11 Evidence You Need to See and Study

Understanding the likely reason for the presence of this hurricane is difficult for those that have not studied enough of the data about events in New York City on the day of 9/11. If you have only reviewed data given out in the official narrative of 9/11, you will not have been exposed to data and evidence which is of vast importance. You will not be aware that this data is so significant and powerful that it was the subject of a "qui tam" case for science fraud case. [278] The defendants were the contractors whom NIST (National Institute of Standards and Technology) employed to contribute to the 10,000 pages of technical reports titled *The Collapse of the World Trade Centre Towers*. This is the US government's official report about the destruction of the WTC towers.

I have written a great deal about 9/11 – or more accurately, I have written about the cover up of what happened, because I was an "insider" in a so-called truth group – which I came to realise was actually working to try and suppress the truth. (See *9/11 Finding the Truth*,[279] if you want to learn many of the "gory" details.)

Dr Judy Wood's *Where Did the Towers Go?* stands as an independent forensic investigation into the destruction of the World Trade Centre. She has stated publicly on many, many occasions [280]:

> "The towers did not burn up, nor did they slam to the ground – they turned mostly to dust in mid air."

"Don't Talk about This Weather..."

On her website, she has a picture[281], which is shown below:

Figure 161 - Dr Wood writes: "My intellectual integrity prevents me from calling this a collapse." [282]

From the evidence Dr Wood examined over approximately seven years, she compiled a list of things that official (and many "alternative") accounts do not explain. Here is a part of that list:

- The Twin Towers were destroyed faster than physics can explain by a free fall speed "collapse."
- They underwent mid-air pulverization and were turned to dust before they hit the ground.
- The protective bathtub, in which the WTC complex was constructed, was not significantly damaged by the destruction of the Twin Towers. (More damage was done to the bathtub by earth-moving equipment during the clean-up process than from the destruction of more than a million tons of buildings above it.)
- The seismic impact was minimal, far too small based on a comparison with the Seattle Kingdome controlled demolition.
- The Twin Towers were destroyed from the top down, not bottom up.

- The demolition of WTC7 was whisper quiet and the seismic signal was not significantly greater than background noise.
- The upper 80 percent, approximately, of each tower was turned into fine dust and did not crash to the earth.
- The upper 90 percent, approximately, of the inside of WTC7 was turned into fine dust and did not crash to the earth.
- One file cabinet with folder dividers survived out of tens of thousands.
- No toilets survived or even recognizable portions of one.
- Evidence that the dust continued to break down and become finer and finer.
- Evidence of molecular dissociation and transmutation, as demonstrated by the near-instant rusting of affected steel.
- Weird fires. The appearance of fire, but without evidence of heating.
- Lack of high heat. Witnesses reported that the initial dust cloud felt cooler than ambient temperatures. No evidence of burned bodies.
- Columns were curled around the vertical axis, where overloaded buckled beams should be bent around the horizontal axis.
- Office paper was densely spread throughout lower Manhattan, unburned, often alongside cars that appeared to be burning.
- Vertical round holes were cut into buildings 4, 5 and 6, and into Liberty Street in front of Bankers Trust, and into Vesey Street in front of WTC6, plus a cylindrical arc was cut into Bankers Trust.
- Approximately 1400 vehicles had to be towed away because they ware 'toasted' in strange ways during the destruction of the Twin Towers.
- The north wing of WTC 4 was left standing, neatly sliced from the main body which virtually disappeared.
- For more than seven years, regions in the ground under where the main body of WTC4 stood continued to fume.
- The WTC1 and WTC2 rubble pile was far too small to account for the total mass of the buildings.
- The WTC7 rubble pile was too small for the total mass of the building and consisted of a lot of mud.
- Eyewitness testimony of Scott-pak (personal air tank) explosions in fire trucks and fire trucks exploding that were parked near the WTC.
- There were many flipped cars in the neighbourhood of the WTC complex near trees with full foliage remaining.

- Magnetometer readings in Alaska recorded abrupt shifts in the earth's magnetic field with each of the events at the WTC on 9/11.

On the following pages, I present a few images from Dr Wood's collection to complement the list above.

What caused the towers to turn to dust?

Figure 162 – WTC Tower turning to dust.

Figure 163 – 70 stories of steel turn to dust...

Why was there so little debris after the destruction?

Figure 164 - WTC site immediately after destruction of WTC 1 and 2

On the afternoon of 9/11/2001 the "rubble pile" left from WTC1 is essentially non-existent. WTC7 can be seen in the distance, revealing the photo was taken before 5:20 PM that day.

How did the inflated tire survive the WTC "plane crash" fireball?

Figure 165 - An official photograph of WTC plane wreckage!

How did this WTC beam get bent into a "horseshoe" shape with no obvious stress, heating or buckling marks?

Figure 166 – WTC beam – deformed.

Why does the car, *parked about half a mile away from the WTC* (on FDR drive) look so burned that the door frame has wilted, yet the rear tyre is still inflated?

Figure 167 – "Toasted" police car

"Don't Talk about This Weather..."

Figure 168 – Remains of a bus just before and after destruction of WTC 7.

These two images were taken before and after the destruction of WTC 7. In the lower image, you can see the remains of WTC 7 which didn't even go the whole way across the street. It appears the bus has moved several feet and become more "mangled" ... What has happened?

"Don't Talk about This Weather ..."

What Turned these Cars Upside Down?

Figure 169 – Inverted car in front of World Financial Centre, September 2001

Figure 170 – inverted police car, September 2001.

What caused this girder in the Banker's Trust/Deutsche Bank building to "crinkle up", when FEMA reported there was no fire in that building?

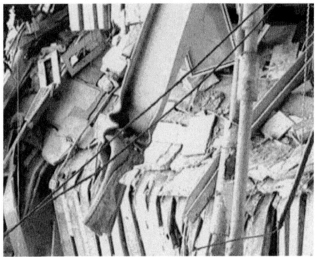

Figure 171 - Inside Banker's Trust/Deutsche Bank building after WTC destruction.

Hurricane Erin on 9/11

It was while researching characteristics of the very fine (nano) dust that was generated in the destruction of the WTC that Dr Judy Wood, former professor of Mechanical Engineering at Clemson University (South Carolina), found the images of Hurricane Erin[283] on satellite photos. Even as I write this in 2017, few people are aware that this category three storm, comparable in diameter to Hurricane Katrina, was closest to New York City at about 8 am on the morning of 9/11/2001. [284] In the morning weather reports, only two out of four local news channels reported the presence of the Hurricane[285] – even though it had been moving, in a fairly straight line, towards New York City since its encounter with Bermuda on 7th September 2001. [286] (This is noteworthy because storm surges would have posed a risk of flooding – potentially to New York's subway system, had the hurricane remained where it was on 9/11/2001. Oddly, on 12 September 2001, the Hurricane made a sharp 135° turn and moved off east, towards Newfoundland.) Dr Wood includes considerable detail about Hurricane Erin in Chapter 18 of her book *Where Did the Towers Go?*

Below, we see one of several traces of the path of Erin:[287]

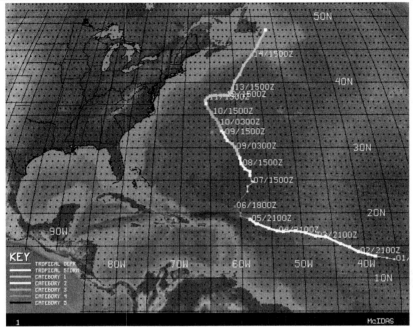

Figure 172 – Path of Hurricane Erin – 01 to 14 September 2001

In her book, Dr Wood notes the details of this track – Hurricane Erin was "born" on about 1 September 2001, and travelled up towards NYC. Hurricane Erin was the closest to NYC on 9/11/2001 and was the largest on this date (although wind speeds were greater the day before). Close-ups from photos of Erin on 9/11 clearly show the plume of material from the destroyed WTC.

The development of Erin is considered, and a comparison made to Hurricane Katrina, since Katrina and Erin were of comparable size (Erin was bigger, by most measures). It is noted that the media reported very little about the potential risk Erin posed around the time of 9/11, compared to what was reported regarding Katrina – even before Katrina made landfall.

Dr. Judy Wood has shown that the barometric pressure and wind speed measured at JFK airport were constant throughout the events of 9/11 – coincident with Erin's closest approach to NYC and greatest expanse of cloud. Morning news reports made light of the hurricane's previous path – while the outer bands of the storm system reached over to Cape Cod. If you *are aware* of this hurricane it is likely that you will read, somewhere, that the hurricane was of no real concern in this case because of a cold front which was advancing from the west. This objection is discussed and addressed in Dr Wood's book. The timing of the arrival of the cold front was also somewhat coincident with the events of 9/119/11 – but this does not change the timing of the movements of the hurricane itself.

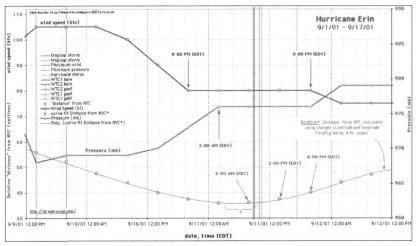

Figure 173 – Chart showing weather and other data during the 24 hours surrounding 11 September 2001

In her book, the development of "super cell" storms is examined and a comparison of their structure to that of a Tesla coil (used to create high voltage electrical discharges) is considered. The possibility is suggested that the electrical properties of large storm systems may have some similarities to those of Tesla coils and that there is a possibility that technology exists to utilise or manipulate the energy in these storm systems for "secondary" purposes.

One of the most striking pieces of the data presented is that from a set of magnetometers monitored by the University of Alaska. [288] Several instruments show significant deviations from "background" or "normal" readings as the events of 9/11 were unfolding. A further selection of this data is presented in relation to variations during the hurricane seasons of 2001, 2004 and 2005.

Readers are invited to compare the media treatments of Hurricane Erin, 2001 to Hurricane Bill (2009) and Hurricane Sandy (2012). The "Erin" name wasn't retired – it was used after 9/11 for other hurricanes, which is against the normal protocol.

What Was the Significance of Hurricane Erin?

The data I have presented above is real. There are two possibilities to consider:

- The presence and movements of Hurricane Erin around the time of 9/11/2001 are coincidence – and the storm had *nothing to do* with anything that happened on that day.
- The presence and movements of Hurricane Erin around the time of 9/11/2001 are *not* coincidence – and the storm *did have something to do* with what happened on that day.

If you were to read Dr Wood's book *Where Did the Towers Go?* you would learn much more about the anomalous effects that have been found in the evidence from 9/11. These effects do show that some type of energy weapon was used to destroy the WTC. The effects that this weapon creates are similar to those created, on a smaller scale, by Canadian researcher and experimenter John Hutchison[289]. It appears that the role of Hurricane Erin was to provide an Electrostatic Field. John Hutchison uses such a field in his experiments. In the 9/11 evidence shown in Dr Wood's book and in John Hutchison's experiments, we can see:

- Lift (levitation) and disruption of samples.
- Material specific effects (metal "melted" with nearby paper unburned)
- Weird fires (fire seen but paper or other materials nearby not burned).
- Transmutation of elements (steel turns to rust quickly)
- Metal turned to jelly – or deformed in ways inconsistent with mechanical stress.

In simple terms, then, it seems Hurricane Erin formed part of an energy interference system – one which was used to turn the WTC towers to dust in about 20 seconds.

As mentioned above, regarding the hurricane itself, considering it had travelled towards New York City for 4 days, it was barely reported. Any memory of its presence was soon all but erased, after talk of Osama Bin Laden, terrorist hijackings, building collapses and the terrible loss of life. No one talked about Erin's sharp change of course on 12 September 2001, because the War Drums were already being sounded.

It appears someone can steer hurricanes – this conclusion has the most profound implications for any discussion of "climate change" or AGW.

Hurricanes Since 9/11

A number of web-based researchers have noted anomalies with hurricanes which have made land fall in the years since 9/11. Hurricane Katrina, which devastated New Orleans made landfall twice and some reports seemed to suggest some damage was intentional – because river levees were "blown" using explosives when the hurricane did seem to cause "the required amount of damage." 290

Hurricane Ophelia's MIMIC (Morphed Integrated Microwave Imagery) trace showed unusual patterns/banding, suggesting some sort of tampering:

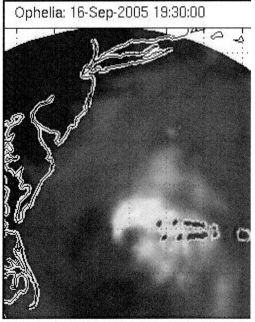

Figure 174 - Hurricane Ophelia - Wisconsin University MIMIC recording.

It is now very difficult to find the original source image for this (an animated GIF file named "KreisbergG2-001.gif"), though a number of copies are posted on several sites – such as Graham Hancock's site. 291 One explanation of the anomalies is as follows: 292

> *Artefacts are commonly associated with features of the image. For instance, in my CT example, the streak artefacts coincide with the metal prosthesis. In ultrasound, you often see echoes of a very sharp boundary because of signal reflections bouncing off other structures. Even Doppler weather radar will*

> *often show echo effects of strong storms. I don't know the specifics of the MIMIC imaging, but since it is radar based, I wouldn't think it unreasonable to see echoes behind an area of strong signal. Additionally, almost every image seems to have artefacts along the boundary where the satellites' images are stitched together. If they were misaligned, and the stitching occurred in a region of high signal, you would expect a line-type artefact.*

The problem here, of course, is that just after these anomalies are seen, the hurricane appeared to change course.

As of finalizing this book, parts of Texas have been devastated by Hurricane Harvey. Some people have asked questions, again, as to how this storm could have strengthened while over land. It is noteworthy how several media articles hurriedly connected the severity of the storm with "global warming and climate change." For example, the UK Guardian proclaimed, "It's a fact: climate change made Hurricane Harvey more deadly." [293] I hope you now have more information that these propaganda-based articles don't ever give you.

CHAPTER FOURTEEN
WHAT FUNDAMENTALIST AGW BELIEFS ARE RESULTING IN ....

"You can't teach an old dogma new tricks."

Usually attributed to Dorothy Parker

Space for Notes Below

YOU Are the Problem!

I have shown you a considerable amount of evidence that you are being lied to about climate change and global warming, and that weather-related 9/11 data and evidence have been hidden from you. Now, we will follow up from earlier commentary about the AGW scam and you can begin to learn (assuming you don't know already) what has been done – largely behind your back – because you have been persuaded that human activity is damaging the environment too much. i.e. you are now being "ordered" to make sacrifices because information has been withheld from you for the benefit of those who truly run the planet

Who wouldn't want to protect the environment and make it "Sustainable"?

As I discussed in Chapter Three, by the early 1990s, global warming propaganda had been underway for several years. It was easy to promulgate this propaganda – showing pictures and videos of polar bears stranded on small blocks of ice, factories belching smoke, polluted lakes and rainforest being chopped down, starving children in Africa and wasteland created by drought. "This has to stop!" we all cried (whilst having no idea what had been kept from us).

Figure 175 – UN Sustainable Development Logo.

Of course, as humans were "destroying the environment," it was time to make human activities "sustainable," and a detailed "plan" of how to do this was presented at the United Nations Conference on Environment & Development in Rio de Janeiro, Brazil, 3 to 14 June 1992. A document called "AGENDA 21 – The United Nations Programme of Action from Rio"[294] was published. The title presumably was meant to be an "Agenda for the 21st century ..." This document is 351 pages long. The logo looks to me like a globe held in a stylised hand, symbolising global control?

Figure 176 – Cover of Agenda 21 document.

A look at the contents page might make you think that it seems like a sensible, benign plan – encouraging careful and responsible use of resources. This, I would agree, is always needed. However, based on what I have learned, I have concluded that this cannot be the primary goal of the plan.

For now, we shall note that even the contents pages contain the word "sustainable" or "sustainability" a total of eight times. Another key section seems to be "4. Changing consumption patterns." Section 4.25 says:

> *Some progress has begun in the use of appropriate economic instruments to influence consumer behaviour. These instruments include environmental charges and taxes, deposit/refund systems, etc. This process should be encouraged in the light of country-specific conditions.*

Since 1992, certainly here in the UK, we have seen a steady increase in systems of "penalties" and "rewards" related to activities which are deemed to have an environmental impact. However, how much environmental damage have these saved? For example, we have solar panels to heat water and generate some electricity. Though this may have saved the "burning of fuels," how much fuel and resources were actually consumed in the manufacture, construction and transport of the solar panels themselves? This would be difficult to quantify accurately.

Section 10 of the Agenda 21 document is called

> "Integrated approach to the planning and management of land resources."

In some areas of the USA, this is being strictly enforced – and there are now some examples [295] of people being charged or arrested for growing food on their own property [296].

Also of note is section 27, "Strengthening the role of non-governmental organizations: partners for sustainable development." One thing about NGOs is that the general public has, in theory, less control over these than government organisations. There are some cases where their funding is less transparent and their connections seem to be dubious. To give an example, Greenpeace can be scrutinized regarding their scale and sources of funding [297]. An article on the "Big Green Radicals" website notes:

> *Undercover reporting by a French journalist under the pen name Olivier Vermont uncovered numerous revealing facts when secretly working at Greenpeace. Vermont spent 10 months working at Greenpeace after presenting himself as an unemployed photographer willing to serve as a volunteer. He ended up serving as an unpaid secretary giving him widespread access to confidential information.*
>
> *In addition to uncovering Greenpeace's collusion with government and industry, his access to secret financial accounts found that a meager six percent of revenue went to field operations while 11 percent went to legal expenses to attack the organization's critics and defend members who had run afoul with the law.*

Of course, the website that this came from mentions nothing about the suppression of things like free energy and ongoing geoengineering activities, which we will discuss later.

Agenda 21 – Random Observations ...

Having glanced through the document, I made the following observations:

- No authors are listed ...
- There is limited referencing.
- Not much reference to carbon footprints specifically (perhaps that buzz phrase came later).

- There is one reference to global warming on page 193, Section 17.125
- There are no graphs, charts or images.

The style is persuasive and the language and arguments fairly simplistic. In materials derived from the Agenda 21 document, you will find certain key phrases and "initiatives." You will find that these are quite common parlance now in local government literature, business and environmental literature and so on. Phrases include:
- Sustainable Development
- Smart Cities
- Smart Growth
- Mega Regions
- Regional Development/Regional Planning
- Environmental justice
- Beyond Traffic
- "Ecology, Equity, Economy"
- Smart Meter
- Smart Motorway

Agenda 21 and the UN – Rosa Koire and "Behind the Green Mask"

Rosa Koire[298] is a forensic commercial real estate appraiser who worked for the California Department of Transportation and her work ultimately brought her into contact with Agenda 21-related documents and procedures. Her twenty-eight-year career in litigation support on land use has culminated in exposing the impacts of Sustainable Development on private property rights and individual liberty in parts of the USA.

In 2005, she was elected to a citizens' oversight committee in Santa Rosa, Northern California, to review a proposed 1,300-acre redevelopment project in which 10,000 people live and work. She, with her partner Kay Tokerud, challenged the fraudulent basis for this huge project (which was in the area she lived).

She has worked diligently to raise awareness of the true goals of Agenda 21 and she has been illustrating to people how it is based on Communitarianism[299] – "a social and political philosophy that emphasizes the importance of community in the functioning of political life." Essentially, Agenda 21 encourages or promotes a form of global communism. In this way of thinking, the rights of the community, in general, outweigh the rights of the individual. On the assumption that all our everyday activities may damage the environment in some way, "permission" is then given to impose restrictions on almost everything we do and how we live. Rosa points out that this form of thinking has gained more acceptance in the corridors of bureaucracy using "the Delphi Technique" where "The goal is to reduce the range of responses and arrive at something closer to expert consensus." [300] This can be done through making people feel awkward if they stand up for a different way of thinking or doing something.

Implemented fairly and equitably, Agenda 21 type policies may improve the quality of life for certain people. However, as Rosa Koire describes in her book in some detail, though it is claimed that changes are made with public consultation, this is not really the case. The way that planning meetings are organised – to essentially give the illusion of public approval for a pre-determined plan, seems to represent a form of global corporatized fascism – with unelected officials and large corporations "calling the shots."

Rosa Koire, like many others who try to reveal truths damaging to those seeking to further the global fascism agenda, has been marginalised, smeared and ridiculed – and just ignored. One example of this can be seen in a Huffington Post interview by Andrew Reinbach[301]. This reporter, (who says he is Jewish in the same interview) tried to link her to "crazy people talking about Zionists." Instead of accurately reporting what Rosa Koire had said to him, he wrote an article[302] which is essentially uncritical of Agenda 21 and, like everyone else who writes about it, he ignores all the reasons why it must be a scam.

Rosa Koire says[303], "It is the blueprint, it is the action plan, to inventory and control all land, all water, all plants, all minerals, all construction, all animals, all means of production, all energy, all information and all the human beings in the world. It is a completely comprehensive plan, it's global and it's implemented locally ... It is in every single town all across the United States and across the world."

In her book *Behind the Green Mask* she walks the reader through various real-world scenarios (based in the USA) of people wishing to use their land in certain ways. She describes how they have been penalised for, or restricted in their plans by Agenda 21 policies which have now "filtered down" to the local level. She points out that "regionalisation" is merely a step to a form of global government.

ICLEI

I·C·L·E·I Local Governments for Sustainability

Figure 177 – ICLEI logo

Rosa has been trying to tell people that Agenda 21 is implemented at the local level via an NGO (Non-Government Organization) called 'International Council for Local Environmental Initiatives', commonly known as ICLEI (pronounced Ik-lee), "Local Governments for Sustainability."

Rosa explains that, "this is paid for by you, the taxpayer, without your consent – because you have not had a vote on its implementation."

On their website[304], we read:

> **Who We Are**
>
> *"By 2050, two-thirds of all humans will be living in cities"*
>
> *ICLEI-Local Governments for Sustainability is the world's leading network of over 1,000 cities, towns and metropolises committed to building a sustainable future.*
>
> *By helping our Members to make their cities and regions sustainable, low-carbon, resilient, ecomobile, biodiverse, resource-efficient and productive, healthy and happy, with a green economy and smart infrastructure, we impact over 20% of the world's urban population.*

Who is "helping" ICLEI …? Well, they have posted an article about the Rockefeller Foundation[305]. I wonder why.

Strong Cities, Smart Cities?

These are promoted as being desirable in Agenda 21/sustainable ways of thinking. Sure, for those of us that live in cities, we want efficient, safe transportation systems and to feel safe walking around these places. However, it seems that Smart and Strong Cities have one thing in common – keeping the citizens in check – under control. For example, extensive usage of surveillance technologies is a given. Smart Cities are being promoted in several ways – one organisation promoting them (itself worthy of further investigation) is Common Purpose: [306]

> "At Common Purpose[307] we believe that cities need to act as magnets. Drawing people in – citizens, leaders and young talent."

WHO believes? Why should anyone care what they believe? We don't even know who "Common Purpose" are!!

Strong Cities are being promoted[308] globally, it seems.

> Launch of Strong Cities Network to Strengthen Community Resilience Against Violent Extremism
> Cities are vital partners in international efforts to build social cohesion and resilience to violent extremism. Local communities and authorities are the most credible and persuasive voices to challenge violent extremism in all of its forms and manifestations in their local contexts. While many cities and local authorities are developing innovative responses to address this challenge, no systematic efforts are in place to share experiences, pool resources and build a community of cities to inspire local action on a global scale.
> "The Strong Cities Network will serve as a vital tool to strengthen capacity-building and improve collaboration," said Attorney General Loretta E. Lynch. "As we continue to counter a range of domestic and global terror threats, this innovative platform will enable cities to learn from one another, to develop best practices and to build social cohesion and community resilience here at home and around the world."

They also have a dedicated website [309] :

> "The Strong Cities Network is the first global network of cities and other sub-national entities working together to build social cohesion and resilience to prevent violent extremism in all its forms."
> "The SCN has issued public statements of support and solidarity for its member cities Beirut, Paris and Kano in light of the recent horrific attacks."
>
> TESTIMONIALS – "I look forward to Aarhus joining the Strong Cities Network steering committee and to speaking about the Aarhus model here in New York. Although radicalization can result from a variety of different causes and be expressed in a

> variety of different ways, there are **a range of trigger factors** that are identical all over the world." –
>
> Jacob Bundsgaard MAYOR OF AARHUS

It seems the "Strong City" agenda is also interlinked with the Green Agenda [310]:

> *"Strong city mayors advance the green agenda"*
>
> *Is it a coincidence that cities with successful green strategies also have strong and successful leadership? Probably not.*
>
> *London, Bristol, New York and Copenhagen all have both. And while strong leadership might not be the only factor driving the green agenda, it is certainly an important one.*
>
> *In December last year, we published our low carbon cities report, which provides examples of how cities in the UK and around the world are going green while supporting economic growth. We found that by providing the vision, strategy and commitment for reducing carbon emissions, mayors such as Bristol's George Ferguson, New York's former mayor Michael Bloomberg and London's Boris Johnson are playing a vital role in advancing this agenda, often ahead of their respective national governments.*

What is the United Nations?

As we mentioned earlier, Agenda 21 was introduced through the United Nations – a name coined by United States President Franklin D. Roosevelt – first used in a declaration on 1 January 1942, during the World War II, when representatives of 26 nations pledged their Governments to continue fighting together against the Axis Powers.

Figure 178 – United Nations logo.

What Fundamentalist AGW Beliefs are Resulting in ....

Figure 179 - UN Conference, 1945

On 24 October 1945, representatives of 50 countries met in San Francisco at the United Nations Conference on International Organization to draw up the United Nations Charter. Those delegates deliberated on the basis of proposals worked out by the representatives of China, the Soviet Union, the United Kingdom and the United States at Dumbarton Oaks, United States in August-October 1944.

The UN headquarters is in New York City (neutral territory, right?). According to the New York City Website[311],

> *The United Nations General Assembly selected New York City as the permanent home of its headquarters in 1946.* **John D. Rockefeller, Jr.** *offered $8.5 million to purchase a six-block tract of land along the East River.*

Of course, we need an organisation promoting World Peace, do we not? Let's have a look at the ideology of this organisation.

Examining the UN's Ideology

One assumes that the UN is a benign organisation, secular and focused on bringing peace and security to the world – it certainly gives that impression. Remember that thing called "The UN Security Council?"

However, how many people know of its connection to the Lucis Trust [312]?

> *"The Lucis Trust is dedicated to the establishment of a new and better way of life for everyone in the world based on the fulfillment of the divine plan for humanity. Its educational activities promote recognition and practice of the spiritual*

> principles and values upon which a stable and interdependent world society may be based. The esoteric philosophy of its founder, Alice Bailey, informs its activities which are offered freely throughout the world in eight languages."

Many people have observed the "Eye in the Triangle" symbology and how it seems to be associated with a kind of occult belief system.

Figure 180 – Lucis Trust logo

One eye, pyramid and triangle symbology seems to be present in the logos of a number of organisations and in various buildings and monuments around the world[313]. (One particularly brazen example is the tomb of Charles Taze Russell in Philadelphia[314] – what has this got to do with a Christian-type burial, I ask). Arguably, we can see a similar symbology in the Lucis Trust logo (consider where the lines seem to cross over, for example).

Again, the Lucis Trust's tagline sounds wonderful *"Let the Plan of Love and Light work out."* What could be wrong with that?

Reading from their own history page[312] (emphasis added):

> The Lucis Trust was established by Alice and Foster Bailey as a vehicle to foster recognition of the universal spiritual principles at the heart of all work to build right relations. ...
>
> A publishing company, initially named Lucifer Publishing Company, was established by Alice and Foster Bailey in the State of New Jersey, USA, in May 1922 to publish the book, Initiation Human and Solar. The ancient myth of Lucifer refers to the angel who brought light to the world, and it is assumed that the name was applied to the publishing company in honour of a journal, which had been edited for a number of years by theosophical founder, HP Blavatsky. It soon became

clear to the Bailey's that some Christian groups have traditionally mistakenly identified Lucifer with Satan, and for this reason the company's name was changed in 1924 to Lucis Publishing Company.

The Arcane School started in 1923, World Goodwill was established in 1932 and Triangles in 1937.
Over the years headquarters offices in New York, London and Geneva have moved locations a number of times. Currently the New York office is located at the southern end of Wall Street in the financial district. In London the office is close to Trafalgar Square in one of the city's most architecturally distinctive buildings, Whitehall Court. And after many years **in a building close to the United Nations**, the Geneva Headquarters is now located on Rue Stand in the business district, close to the lake.

Figure 181 – Alice Bailey

A few years after the formation of the UN, Eleanor Roosevelt on June 2, 1952 – World Invocation Day[315] – read a short "prayer" called the Great Invocation [316] (emphasis added).

The Great Invocation
From the point of Light within the Mind of God
Let light stream forth into the minds of men.
Let Light descend on Earth.
From the point of Love within the Heart of God
Let love stream forth into the hearts of men.
May Christ* return to Earth.
From the centre where the Will of God is known
Let purpose guide the little wills of men –
The purpose which the Masters know and serve.
From the centre which we call the race of men
Let the Plan of Love and Light work out

> And may it seal the door where evil dwells.
> Let Light and Love and Power restore the Plan on Earth.

Figure 182 – Eleanor Roosevelt

And who are the "Masters" mentioned here? The "prayer" or "invocation" is accompanied by a kind of "disclaimer" which reads as follows: "Many religions believe in a World Teacher, a "Coming One", knowing him under such names as the Lord Maitreya, the Imam Mahdi, the Kalki Avatar and the Bodhisattva. These terms are sometimes used in versions of the Great Invocation for people of specific faiths."

Lucis Trust and ... Astrology ... in 2015?

Again, from the Trust's own website ...[317]

> *Sustainable Development Goals, the United Nations and the Libra Full Moon*
>
> *A significant event in the heavens occurred during the period of the Sustainable Development Summit at UN Headquarters in New York. The full moon (with the sun in Libra) was late in the evening of Sunday, the final day of the Summit. And it was not just any full moon – popularly called a 'Supermoon' the moon was especially close to earth so it appeared much larger in the night sky than usual. And there was a full lunar eclipse producing a red effect on the moon. There have been only 4 full eclipses during a supermoon since 1900 – so this was an unusual astronomical event. It should not surprise us that a significant alignment in the heavens was accompanied by such a significant gathering at the UN.*

So now they're mixing in astrology too ... and remember this is an organisation connected to a body which is supposedly helping to run the planet ...

In a similarly unusual way, they had a keynote tagline for their 2016 Arcane School Conference[318]:

> Let the Group strive towards that subjective synthesis and telepathic interplay which will eventually annihilate time.

Lawrence Newey, one of the speakers at this conference talked about forms of energy such as Cold Fusion (LENR) in a talk titled "Discipleship and the New Fire." [319] The speaker even talks about energy being drawn down from the "Etheric Plane." Doesn't this sound similar to what Wilhelm Reich and Trevor James Constable were talking about – some form of "ether energy" ...? Yet keep in mind, we are discussing an organisation that is closely connected to the United Nations.

Climate Change, Extreme Weather, Floods etc

In 2015, we again saw awful flooding in parts of the UK[320] – such as Carlisle, Lancaster, Hebden Bridge, York, parts of Leeds and elsewhere. Significant flooding was also seen in the UK in 2007 (Hull[321]), 2009 (Cockermouth and elsewhere in Cumbria[322]), 2012 (Exeter area and Malmsbury), 2014 (Dawlish and Somerset Levels). In 2015, they started naming storm systems in the UK for the first time[323] (Desmond, Eva etc).

Figure 183 – UK flooding

Who would consider that these events were engineered ...? Who would consider that this was an example of the Hegelian Dialectic at work again – but on a massive scale? Agenda 21's success relies on *people NOT thinking of connections. They must NOT entertain the idea of control and deception of this magnitude ...*

EU Natura Policy 6 – Allowing Floods

In 2014 much upheaval, destruction and disruption was apparent in Somerset[324], UK. Some people quickly became aware of one of the possible contributing causes of this widespread flood damage – the fact that dredging of rivers had been halted in some areas. From Christopher Booker in the *Daily Telegraph*:

> "These have included the EU's Natura 2000 strategy along with a sheaf of directives on "habitats", "birds", "water", and not least the "floods" directive of 2007, which specifically requires certain "floodplains" to be allowed to flood. In 2008, when the EA was run by Baroness Young, this was reflected in a policy document which classified areas at risk of flooding under six categories, ranging from those in "Policy Option 1", where flood defences were a priority, down to "Policy 6" where, to promote "biodiversity", the strategy should be to "increase flooding." The Somerset Levels were covered by Policy 6."

Figure 184 – Flooding in Somerset Levels, UK – 2014

In "Policy 6" areas, dredging was stopped – based on an EU directive – itself seemingly part of "Agenda 21 land management" guidelines or policies. And, indeed, from an EU Natura 2000 Webpage[325] we can read:

> Natura 2000 is not a system of strict nature reserves from which all human activities would be excluded. While it includes strictly protected nature reserves, most of the land remains privately owned. The approach to conservation and sustainable use of the Natura 2000 areas is much wider, largely centred on people working with nature rather than against it. However, Member States must ensure that the sites are managed in a sustainable manner, both ecologically and economically.

But who would consider that the climate/weather itself could also be being engineered to make such events happen …? The Hegelian Dialectic can be used to guide people away from thinking about such a conclusion and make them focus on more prosaic explanations, whilst also shouting things like "something should be done by government to protect our property and livelihood!" (For example, build up flood defences and take steps to somehow reduce insurance premiums etc.)

Fracking is Fine … Really!

If there has been a push towards "Sustainable Living" since about 1992, then how is it that licenses for fracking have been granted around the world? Is it not really the case, then, that large corporations are exempt from "sustainability" rulings/guidelines, yet individuals acting on a much smaller scale are not exempt?

Figure 185 – Cuadrilla – fracking site.

Why is it that police forces here in the UK have been employed to protect corporate interests[326] rather than environmental ones – or the interests of residents in the affected communities[327]? I have concluded that a study of what has happened at fracking sites, and the way protestors have been treated proves that Agenda 21 is not what is claimed – and *is* much more likely a plan to strengthen corporate control of land, resources and people.

The Hegelian Dialectic here then can be seen in the conflict between environmentalist groups and another group who advocate the development and use of natural resources, to benefit the economy of a nation, county or area. As you will read below, I know there is no need to frack anything anywhere. It's another "phoney bone of contention" which no one that I have heard talking about fracking or Agenda 21 will discuss with the same candour.

COP21 – The Marriage of the "Terrorism" and Climate Change Scams?

The "Paris Climate Conference – COP21 – Conference of Parties" took place between 30 November 2015 and 12 December 2015. From their website[328] we read:

Figure 186 – Climate change activists in Paris in 2015

> "The international political response to climate change began at the Rio Earth Summit in 1992, where the 'Rio Convention' included the adoption of the UN Framework on Climate Change (UNFCCC). This convention set out a framework for action aimed at stabilising atmospheric concentrations of greenhouse gases (GHGs) to avoid "dangerous anthropogenic interference with the climate system …. The UNFCCC which entered into force on 21 March 1994, now has a near-universal membership of 195 parties."

Just before the conference 13/14 November 2015 saw the alleged bombings and the deaths of over 100 people[329]. Paris then went into "Lockdown." However, terrorist incident drills were going on at the same time as the events happened[330] – just like they were on 7th of July 2005[331] – the date of the alleged bombings in London.

I suggest that, to notch up the "fear and trauma programming," and reduce any effects that scepticism might be having on the progress of the global governance and control agenda, what we experienced was the "interlocking" of two global psychological operations – in a major world capital. This was then blasted 24/7 across the world's media ... to sear away any dissent and reinforce the previous decades' long programming. An immediate result was that "climate change" protestors were arrested[332] (though the banners they held almost certainly did not show any of the more challenging evidence covered in this book.)

CHAPTER FIFTEEN AN ANSWER TO ENVIRONMENTAL POLLUTION AND DESTRUCTION?

"Those who deny freedom to others, deserve it not for themselves."

Abraham Lincoln (1809-65), Speech, 19 May 1856

Space for Notes Below

Revolutionary Energy Technologies

Over the years, following the footsteps of many people who are cleverer, more resourceful and more resilient than me, I have come to know that we have been deprived of world-changing technologies and techniques which, at a stroke, render any arguments about scarcity of resources or worries about the environment as next to meaningless. The same dark forces that are pushing Agenda 21, and similar programmes, are the most likely culprits for suppressing technologies. In this short chapter, I will list some examples of projects and technologies which could have yielded cheap and/or pollution-free energy sources which could be used to mitigate any or all of the perceived problems related to alleged climate change or global warming. I leave the reader to pursue the details.

"Salter's Duck" – Killed off in the 1980s

Stephen Salter has drawn some attention with his "cloud maker" idea[333]. However, he does not seem to talk about his "duck" (sea) wave-energy device which could have produced large quantities of energy and was reviewed independently and funding recommended. Funding for the Salter's project was then withdrawn without the agreement of the independent reviewer.[334,335]

Meyer Cell – Efficient Water Splitter – Emission-less Car

Stanley Meyer developed this extremely efficient (unconventional) electrolyser – and was eventually poisoned following a meeting with investors.[336]

Safe Hydride Storage System for Hydrogen Car – Illegal

Bob Lazar's Hydrogen Car System[337] is an example of a fairly conventional alternative system for running a car on Hydrogen Gas[338]. In Lazar's design, hydrogen is stored in a metal hydride material and is released on demand, while the car is being drive. It gets around the hydrogen storage problem. It has been suppressed,[339] because one of the materials that is used in its construction, a certain hydride, is a "weapons grade material" – so it cannot be sold.[340] This is true even though the material itself is safe.

Figure 187 – Bob Lazar's hydrogen generator and storage system for cars.

Lazar was told he had to comply with an ever-increasing list of "health and safety" issues. The system works, but has never been approved, and no one has stepped forward to help Lazar overcome any remaining issues.

Cold Fusion/LENR – Killed off by Vested Energy Interests

Dr Eugene Mallove was the President of the non-profit New Energy Foundation, Inc. He held a Master of Science Degree and Bachelor of Science Degree in Aeronautical and Astronautical Engineering from MIT, and a Science Doctorate in Environmental Sciences from Harvard University. He worked for high technology engineering companies such as Hughes Research Laboratories, TASC (The Analytic Science Corporation), and MIT Lincoln Laboratory.

He was Chief Science Writer at the MIT News Office when discovery of a new energy phenomenon, incorrectly dubbed "Cold Fusion," was announced by electrochemists Stanley Pons and Martin Fleischmann. Mallove came to understand that current physics, based on Quantum Mechanics and Relativity, is almost completely wrong. He also became interested in Wilhelm Reich's work following his examination of working Orgone Accumulators.

Figure 188 – from left to right – Dr Eugene Mallove, Dr Stanley Pons and Dr Martin Fleischmann.

In 1999, Mallove produced a documentary called "Fire From Water" in which scientists who worked on Cold Fusion Experiments were interviewed.[341] Mallove was murdered in May 2004.[342]

Dr Stanley Pons and Dr Martin Fleischmann were well-regarded chemists at the University of Utah. They had worked for five years to develop a process which seemed to yield too much energy to be a simple chemical reaction. They stated they had observed evidence of nuclear reactions – such as neutron emission, transmutation and "excess heat." In 1989 they announced their findings at a press conference. They were criticised for this as they would normally have "announced" their research in a paper which would be peer reviewed – in a scientific journal. Cold Fusion should more correctly be called Low Energy Nuclear Reactions (LENR). The somewhat inaccurate description "Cold Fusion" appears to have been coined by Dr Steven E "Thermite" Jones – whom I wrote more about in my free eBook *9/11 Finding the Truth*.

It seems the whole cold-fusion affair was managed from start to finish. The goal was to trash and debunk the science, and it seems a whole array of sceptics and debunkers were "wheeled out" to "stomp out the fire." It was the first time the phrase "pathological science" was brought into general use. People I speak to today say things like "their results have never been reproduced!" (which is wholly untrue). Basically, in many cases, it is difficult to reliably reproduce the conditions under which the reaction will take place. Different reproductions have different results in terms of seeing:
- Excess heat
- Evidence of a nuclear process (neutrons and or tritium)
- Transmutation of elements

Research is ongoing to this day – interested readers should check out the Martin Fleischmann Memorial Project.

Other notable examples of energy technology

These include:

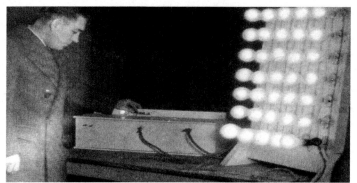

Figure 189 – Demonstration of Dr T. Henry Moray's Radiant Energy Device

- Certain elements of Nikola Tesla's work, such as his Wardenclyffe Tower Project.
- Dr T. Henry Moray's Radiant Energy Device[343] (1920s onwards)
- Stan Meyer's Water Fuel Cell[344]
- Lutec Device – Australia (c. 2000)
- Bruce De Palma's N Machine and derivatives.

Investment in Energy Alternatives?

The real problems are much, much different – and deeper and wider – than those assumed to be caused by CO_2 emissions (or any other emissions for that matter). Traditional renewable alternatives such as wind turbines, solar heating, solar photovoltaic systems and geothermal energy systems are being utilised more – as some technologies that are publicly available become cheaper and/or more robust, but all these systems have limitations because of things like energy storage and maximum energy output for a given situation (e.g. solar PV is much less useful in northern latitudes). They also are costly to build.

Know the True Situation

It is now time for a fundamental change in thinking and a realization of "how things really are." From other research, it becomes clear that much less polluting – even non-polluting – and perhaps limitless energy sources have been perniciously (and fatally) suppressed.

Additionally, the evidence from 9/11, amassed in Dr Judy Wood's book, proves that some type of "free energy" technology has already been weaponised and has been used. This proves it is real. It also proves that the "worst people" already have access to this technology.

Where will you go from here …? Back to supporting AGW lies and measures to supposedly mitigate this and make your life more reasonable? Perhaps it would be better if people demanded more truth and accountability, based on now easily accessible facts.

CONCLUSIONS

"Truth has no special time of its own. Its hour is now – always."

Albert Schweitzer (1875–1965) *Out of My Life and Thought*

Space for Notes Below

Remember the Warning?

In Chapter Two, I referred you to President Dwight Eisenhower's warning about the Military Industrial Complex. I have given you a glimpse at some of the real events at the World Trade Centre in New York – where three towers were "dustified" by an undisclosed technology. What I didn't tell you is that a company called Science Applications International Corporation (a nice "generic sounding" name) was involved in emergency response operations at "Ground Zero" on 13 September 2001 and they also participated in the compilation of the NIST technical reports about the destruction of the WTC itself. SAIC is a large Defence Contractor. Denise McKenzie, a whistle-blower in Dr Steven Greer's Disclosure Project stated in March 2001 that SAIC are heavily involved in classified, secret or "black" programmes. [345] We know that the military were involved in weather modification operations in Vietnam in the 1960s and 1970s. This has been disclosed.

Let us accept, for the moment, that what Wilhelm Reich did in the 1950s and what Trevor James Constable and James De Meo did in the 1980s and 1990s was all real. They did these things with a budget of a few thousand dollars and some ingenuity and hard work.

Now, what do you think is possible with almost unlimited funds over a period of say, 50 years – working in protected, secure and completely hidden development environments, with a global propaganda network and a squadron of compliant media mouthpieces and officials at your disposal …?

The Global Deception Grid

Agenda 21 represents some of the "construction material" for the Global Grid of Deception. I argue that two of these elements of "the global deception grid" interlocked in Paris in November 2015 – with the supposed terrorist attack near the time of the "climate change" summit there.

As you should know by now, energy usage and climate change are never discussed in a way which considers all the evidence I present here. For example, it is never, ever discussed in any mainstream source, nor even in any popular *alternative* media outlet, that some group has the technology that can turn three buildings to dust in 10 seconds or less each. It is never discussed that some group can steer hurricanes and use their electric field as a component of a weapon system of some kind[283] – and that this weapon system was used on 9/11 to destroy most of the WTC complex.[346]

Use of energy is one of the keys to global control of the masses. If free energy were available for everyone, it would change everything. The worst people already have access to this technology – and they used it on 9/11. They have successfully covered this up.

Obviously, organisations such as Amnesty International, CND, Stop the War will have none of this. There is too much inertia, infiltration and control of NGO's who pride themselves on being "pro-peace" and "defending human rights." And in any case, several of them have some kind of connection either to the UN itself, the Tavistock Institute, or other groups with seemingly dubious agendas. (See Daniel Estulin's book *Tavistock Institute: Social Engineering the Masses*.)

Ongoing fear and confusion is required to prevent people from thinking for themselves and exploring the evidence. Researchers who talk about Agenda 21 and oppose it (mainly for the reasons that it places restrictions on personal freedom) don't seem to know that any basis for Agenda 21 is completely and comprehensively undermined by a knowledge of what really happened on 9/11.

So, the set of "fear cards" dealt to you by mainstream sources which help structure the global control grid include:

- Terrorist attacks
- Climate change/extreme weather
- Biowarfare or virulent disease outbreak
- Cancer
- Fuel/Energy shortage
- Food crisis
- Economic collapse

Agenda 21 can be implemented because the perception is created that you would then be "protected" in "smart cities" functioning on ideals of "Social Equity," while other areas are "left wild" and by and large, you wouldn't be able or allowed to go there.

Cover Story and Real Story

As implied elsewhere in this book, there is some evidence relating to the "climate change" issue which seems to be a "no go area" for almost everyone. We have a number of speakers who challenge the CO_2 lie, and some have been featured in the mainstream such as Dr Tim Ball, Piers Corbyn, Johnny Ball and many others who, to my knowledge, have never mentioned the movements of Hurricane Erin – even though this would tend to support their conclusion that the CO_2 story is a lie. In a similar way, they have persuaded most people that, for example, the appearance of many aircraft trails are quite normal. They have started to acclimatize people to the idea of geoengineering – so that, at some point in the future, you or your children will accept it as a normal, necessary activity. For those that haven't been taken in by all the propaganda, they have encouraged talk of toxic spraying programmes and HAARP to confuse the issue and divert your attention to an area which is of far less concern to them.

I have often wondered if the trails we see from aircraft are actually appearing because the atmosphere is being externally manipulated – not because aircraft are emitting toxic spray.[347] As I alluded to above, since the 1950s, I think that those running the planet have perfected technology which Reich originally developed – to manipulate the weather and the atmosphere. Therefore, we experience some days where there are no trails to be seen at all – the energy manipulation is changed from day to day and hour to hour and region to region. (I don't think this is done with HAARP, as many suggest. I suggest the weather control technology is something which operates in a significantly different manner to HAARP.)

Even popular (though all-but anonymous) "YouTubers" Dutchsinse[348] and WeatherWar101[349] failed to mention or give a full picture of the significance of Hurricane Erin on 9/11, even though the information has been available since 2008. This is quite staggering – seeing as the data about Hurricane Erin and its relationship to the events of 9/11 is the most important meteorological data in, well, probably ever …

What seems to have happened then is that most of the western world has been persuaded to believe a false story about energy scarcity, that humans can only cause irreparable environmental damage and destruction. All the while, the elite groups that run the planet know about free energy technologies. They also know how to control the weather – and they are already doing this – but not in a way that is blatantly obvious. They have sold you a fake story about CO_2 and global warming and press-ganged you into agreeing to measures which curb your freedom. They set up a decades-long propaganda programme – such that those who write the sorts of things I have become hated or despised and the subject of ridicule, smear campaigns or other emotionally-driven reactions. "Thought policing" when dealing with these issues (and similar ones) has become the norm, without people even noticing.

The "emotional plague" that Wilhelm Reich spoke of is real – and it has infected a very large percentage of the population – particularly those in positions of authority of one kind or another. Those affected by the plague have lost their powers of unbiased observation and are unable to entertain the conclusion that a secret group of some kind is doing things without their knowledge or consent. Every denial of evidence and every repetition of their propaganda that is made, without employing critical thought, gives more power to people that operate "behind a veil" of secrecy and deception. People affected by the emotional plague do this, in many cases, without realising what they are doing – as if certain parts of their minds are "closed off" or "straight-jacketed." In many cases, they make a living because they have fallen for one part of this deception or another. The system seems to have been designed that way ...

The truth seems to be, that the people running things need to control all aspects of your life – for reasons they won't tell you (and reasons I can only guess at). They are desperate to do this – and desperate to keep your mind in a state where you would normally reject what I am telling you. Somehow, despite their best efforts, you have broken through all the nonsense and managed to read this page... Isn't it amazing?

It's not us – It's someone else

Hence, I hope you now will consider how important it is to realise that in any arguments about "humans damaging the environment" – where people say things like:

> "We need to change our ways – we're destroying the planet ..."

IT'S NOT US who are really responsible for this at all. (By "us" I mean the ordinary people just trying to live their lives and help one another in some way.) Ordinary people are NOT TO BLAME for being forbidden from having access to knowledge and technology to live in harmony with the rest of the planet. ANOTHER GROUP has kept secrets and told lies to protect itself and hide itself. *This "other group" has technology and they've kept it to themselves and made (most of) the rest of us ignorant of it.* They turned it into a weapon and destroyed the World Trade Centre with that weapon. It also seems very much like they are controlling the weather with similar technology [350]. That is the *true situation* which few will properly investigate or articulate – because they haven't seen the relevant available evidence. I hope you can and will study the evidence if you're not already aware of it.

The weather modification technology *must be disclosed*, exposed before it could ever be used properly. This book is an attempt to disclose this situation to you. How many people will pay attention?

If this book has not made you in some way "angry" then it is hoped it has made you curious to learn more – perhaps, through your curiosity, you can then help to change the world for the better.

At this stage, all we can do is make more people aware of the information presented here and challenge those who should be looking into it (GO's and NGOs alike). In saying this, however, these "challenges" probably should be done in a "one on one" or face-to-face setting. In my experience, writing emails/letters to MPs, government departments or environmental or other organisations does not work. You are just stonewalled and not heard.

Am I right? If you agree, please do what you can to share your new-found knowledge with anyone who will listen – but especially those who could make a difference.

REFERENCES

1 http://www.realclimate.org/index.php/archives/2007/04/the-lag-between-temp-and-co2/
2 https://wattsupwiththat.com/2012/07/23/new-research-in-antarctica-shows-co2-follows-temperature-by-a-few-hundred-years-at-most/https://wattsupwiththat.com/2012/07/23/new-research-in-antarctica-shows-co2-follows-temperature-by-a-few-hundred-years-at-most/
3 http://www.skepticalscience.com/about.shtml
4 http://www.harc.edu/AirQualityClimate/HoustonUrbanHeatIslandEffect/tabid/301/Default.aspx
5 http://climatechange.sea.ca/kyoto_protocol.html
6 http://unfccc.int/meetings/cop_15/items/5257.php
7 http://www.ourdocuments.gov/doc.php?flash=old&doc=90&page=transcript
8 https://www.rand.org/topics/delphi-method.html
9 http://www.crossroad.to/articles2/05/dialectic.htm
10 http://www.simplypsychology.org/asch-conformity.html
11 https://www.youtube.com/watch?v=TYIh4MkcfJA
12 https://www.youtube.com/watch?v=Uz0HzS1O-ug
13 https://books.google.co.uk/books?id=R9-3AAAAIAAJ&source=gbs_book_other_versions
14 https://youtu.be/hYUIwdqCpU0?t=27m26s
15 http://www.digitalspy.com/tv/feature/a581066/edge-of-darkness-does-bbc-twos-beloved-drama-stand-test-of-time/
16 http://worldnews.nbcnews.com/_news/2012/04/23/11144098-gaia-scientist-james-lovelock-i-was-alarmist-about-climate-change
17 http://ecolo.org/lovelock/lovedeten.htm
18 http://www.telegraph.co.uk/news/earth/paris-climate-change-conference/12035401/Farewell-to-the-man-who-invented-climate-change.html
19 http://mauricestrong.net/index.php/tributes5/113-strong-lubbers
20 http://www.un.org/geninfo/bp/enviro.html
21 http://www.youtube.com/watch?v=wIl2gdDtbCg&NR=1
22 http://www.foxnews.com/entertainment/2010/06/24/paul-mccartney-global-warming-holocaust-deniers/
23 http://web.archive.org/web/20121226083318/http://www.uni-graz.at/richard.parncutt/climatechange.html
24 http://web.archive.org/web/20130125091913/http://www.uni-graz.at/richard.parncutt/climatechange.html
25 http://www.dailymail.co.uk/news/article-1199104/Peter-Sissons-BBC-standards-falling--bosses-scared-it.html#ixzz0LDzY3uNe&C
26 http://www.climatechangefacts.info/ClimateChangeDocuments/LandseaResignationLetterFromIPCC.htm
27 http://www.bbc.co.uk/wear/content/articles/2007/03/27/climate_countdown_david_bellamy_feature.shtml
28 http://www.telegraph.co.uk/culture/tvandradio/9817181/David-Bellamy-tells-of-moment-he-was-frozen-out-of-BBC.html
29 http://www.dailymail.co.uk/sciencetech/article-1359350/Zoe-Balls-father-Johnny-vilified-questioning-global-warming.html
30 http://www.dailymail.co.uk/news/article-3355441/QUENTIN-LETTS-vaporised-BBC-s-Green-Gestapo.html
31 http://www.bbc.co.uk/programmes/b0641815
32 http://www.telegraph.co.uk/news/worldnews/europe/france/11931645/Frances-top-weatherman-sparks-storm-over-book-questioning-climate-change.html
33 http://www.wsj.com/articles/SB10001424052702303480304579578462813553136 https://www.youtube.com/watch?v=4LkMweOVOOI
34 https://www.youtube.com/watch?v=4LkMweOVOOI
35 http://epw.senate.gov/public/index.cfm?FuseAction=Minority.Blogs&ContentRecord_id=b35c36a3-802a-23ad-46ec-6880767e7966
36 http://zwww.cato.org/special/climatechange/alternate_version.html
37 http://australianconservative.com/2010/10/michael-coren-with-dr-tim-ball/
38 http://blogs.telegraph.co.uk/news/jamesdelingpole/100058265/us-physics-professor-global-warming-is-the-greatest-and-most-successful-pseudoscientific-fraud-i-have-seen-in-my-long-life/
39 https://www.desmogblog.com/christopher-monckton

References

[40] https://www.desmogblog.com/richard-lindzen
[41] http://www.exxonsecrets.org/html/orgfactsheet.php?id=21
[42] http://www.petitionproject.org/instructions_for_signing_petition.php
[43] http://www.petitionproject.org/review_article.php
[44] http://www.odu.edu/ao/instadv/quest/Greenhouse.html
[45] http://theconsensusproject.com/
[46] http://www.cs.uni.edu/~okane/warming.html
[47] http://www.pnas.org/content/99/17/10976.full?maxtoshow=&HITS=10&hits=10&RESULTFORMAT=&fulltext=genesis+of+hydrocarbons+and+the+origin+of+petroleum&searchid=1085470440708_510&stored_search=&FIRSTINDEX=0
[48] https://www.youtube.com/watch?v=2cUg3lDgJ20
[49] http://pubs.acs.org/doi/abs/10.1021/ed031p326
[50] https://www.theguardian.com/politics/2003/sep/06/september11.iraq
[51] http://junkscience.com/FOIA/
[52] https://www.skepticalscience.com/broken-hockey-stick.htm
[53] https://www.amazon.co.uk/Illusion-Climategate-Corruption-Science-Independent-x/dp/1906768358
[54] https://books.google.com/ngrams/graph?content=geoengineering&year_start=1800&year_end=2000&corpus=15&smoothing=3&share=&direct_url=t1%3B%2Cgeoengineering%3B%2Cc0#t1%3B%2Cgeoengineering%3B%2Cc1
[55] http://royalsociety.org/policy/publications/2009/geoengineering-climate/
[56] http://royalsociety.org/Geoengineering-taking-control-of-our-planets-climate/
[57] http://aparc.stanford.edu/publications/the_geoengineering_option/
[58] http://news.bbc.co.uk/1/hi/programmes/6354759.stm
[59] http://www.whatdotheyknow.com/request/chemtrails_10
[60] http://royalsociety.org/news/stop-co2/
[61] http://www.cspg.org/documents/Conventions/Archives/Gussow/2008Gussow/presentations/021-Climate_and_Carbon_Engineering.pdf
[62] http://www.publications.parliament.uk/pa/cm200910/cmselect/cmsctech/221/221.pdf
[63] http:/www.publications.parliament.uk/pa/cm200910/cmselect/cmsctech/221/221.pdf
[64] https://www.gov.uk/government/uploads/system/uploads/attachment_data/file/47928/569-gov-response-commons-science-tech-5th.pdf
[65] http://www2.eng.cam.ac.uk/~hemh/SPICE/SPICE_announcement_14_Sept_2012.pdf
[66] http://www.guardian.co.uk/environment/2012/feb/09/at-war-over-geoengineering
[67] http://www.homepages.ed.ac.uk/shs/Climatechange/Geo-politics/Victor_et_al_ForeignAffairs2009.pdf
[68] http://www.marcgunther.com/is-geoengineering-inevitable/
[69] https://www.congress.gov/bill/109th-congress/senate-bill/517
[70] http://www.un-documents.net/a31r72.htm
[71] https://www.congress.gov/bill/107th-congress/house-bill/2977/text
[72] https://royalsociety.org/events/2010/geoengineering/
[73] http://www.cquestrate.com/the-idea/detailed-description-of-the-idea
[74] https://www.populationmatters.org/porritt-population/
[75] http://www.crispintickell.com/page109.html
[76] http://downloads.royalsociety.org/audio/DM/DM2010_09/PANEL DISCUSSION.mp3
[77] http://www.checktheevidence.com/pdf/chemtrails -leaflet black and white.pdf
[78] https://royalsociety.org/events/2013/climatescience-next-steps/
[79] http://www.weatheraction.com/
[80] https://www.e3g.org/about
[81] https://www.e3g.org/showcase/degrees-of-risk/
[82] https://www.e3g.org/people/john-ashton
[83] https://www.cfr.org/event/conversation-us-secretary-state-hillary-rodham-clinton-1
[84] https://www.cfr.org/councilofcouncils/global_memos/p32983
[85] https://www.cfr.org/content/about/annual_report/ar_2008/Studies2008.pdf
[86] https://www.nasa.gov/centers/goddard/news/topstory/2003/0313irradiance.html
[87] http://earthsky.org/space/what-are-coronal-mass-ejections
[88] http://www.aharfield.co.uk/lightning-protection-services/about-lightning
[89] https://www.britannica.com/topic/Maunder-minimum
[90] http://www.odlt.org/dcd/docs/john_eddy_Maunder_Minimum.pdf
[91] http://physicsworld.com/cws/article/news/1998/nov/26/cern-plans-global-warming-experiment
[92] http://www.nationalpost.com/news/story.html?id=975f250d-ca5d-4f40-b687-a1672ed1f684
[93] http://pages.science-skeptical.de/MWP/MedievalWarmPeriod.html
[94] http://scienceandpublicpolicy.org/images/stories/papers/monckton/monckton_what_hockey_stick.pdf

[95] https://www.britannica.com/science/medieval-warm-period
[96] http://www.johnstonsarchive.net/environment/warmingplanets.html
[97] https://www.sciencedaily.com/releases/1999/02/990223083118.htm
[98] https://www.hq.nasa.gov/office/pao/History/presrep95/solarsys.htm
[99] http://physicsworld.com/cws/article/news/2001/jan/18/night-time-on-venus
[100] https://en.wikipedia.org/wiki/File:Helkivad_%C3%96%C3%B6pilved_(2).jpg
[101] http://hubblesite.org/news_release/news/1995-16
[102] http://hubblesite.org/image/482/news_release/1997-15
[103] http://www.berkeley.edu/news/media/releases/2004/04/21_jupiter.shtml
[104] http://science.nasa.gov/science-news/science-at-nasa/2006/02mar_redjr/
[105] http://hubblesite.org/newscenter/archive/releases/2008/23/image/a/
[106] http://science.nasa.gov/science-news/science-at-nasa/2010/20may_loststripe/
[107] http://web.wellesley.edu/PublicAffairs/Releases/2003/060403.html
[108] http://www.newscientist.com/article/dn9100-saturns-rotation-puts-astronomers-in-a-spin.html
[109] https://www.newscientist.com/article/dn9100-saturns-rotation-puts-astronomers-in-a-spin/
[110] http://news.discovery.com/space/saturns-north-pole-hexagon-mystery-solved.html
[111] http://www.planetary.org/blog/article/00002471/
[112] http://www.spaceweather.com/swpod2010/28dec10/wesley1.jpg?PHPSESSID=fsjl7fe642n28prenc5k3ga432
[113] http://www.spaceweather.com/archive.php?view=1&day=28&month=12&year=2010
[114] http://abyss.uoregon.edu/~js/images/uranus_hst.gif
[115] http://science.nasa.gov/science-news/science-at-nasa/1999/ast29mar99_1/
[116] http://news.nationalgeographic.com/news/2011/11/111021-uranus-planet-new-spot-storm-methane-gemini-space-science/
[117] http://tellus.ssec.wisc.edu/outreach/neptune.htm
[118] http://www.news.wisc.edu/newsphotos/neptune.html
[119] http://news.mit.edu/2002/pluto
[120] https://www.space.com/3159-global-warming-pluto-puzzles-scientists.html
[121] https://www.britannica.com/topic/interstellar-medium
[122] http://ruk.usc.edu/bio/dons/ds_biosk.html
[123] http://science.nasa.gov/science-news/science-at-nasa/2003/12nov_haywire/
[124] http://www.holoscience.com/synopsis.php
[125] http://www.holoscience.com/news.php?article=8pjd9xpp
[126] http://web.archive.org/web/20041126100403/http://www.es.lancs.ac.uk:80/Hazelrigg/amy/Observation Pages/Contrail Types.htm
[127] https://science-edu.larc.nasa.gov/contrail-edu/science.php
[128] https://www.metabunk.org/threads/exclusive-leaked-photos-of-chemtrail-dispersal-system.2772/
[129] http://www.reallibertymedia.com/2013/11/exclusive-photos-of-chemtrail-dispersal-system/
[130] https://www.carnicominstitute.org/articles/bio1.htm
[131] https://www.chemtrailsprojectuk.com/evidence/rainwater-test-kit-results-map/
[132] https://www.metabunk.org/the-claims-of-francis-mangels-a-factual-examination.t154/
[133] http://ir.uiowa.edu/cgi/viewcontent.cgi?article=1076&context=igsar
[134] http://www.agriculturedefensecoalition.org/content/about-rosalind-peterson
[135] https://www.youtube.com/watch?v=U9NKvT7Iuhg
[136] https://www.metabunk.org/debunked-rosalind-peterson-leaker-addressing-un-about-chemtrails-and-geoengineering.t3514/
[137] https://climateviewer.com/chemtrails/
[138] https://wn.com/history_of_uk_chemical_spraying_admitted_by_bbc_documentary_chemtrails
[139] http://politics.guardian.co.uk/news/story/0,9174,688098,00.html
[140] http://web.archive.org/web/20040612194433/http:/ju2003.pnl.gov/pdfs/J-URBAN-Q&A-sheet6-10-03.pdf
[141] http://contrailscience.com/
[142] https://www.metabunk.org/about-metabunk.t1966/
[143] http://www.orgonelab.org/chemtrails.htm
[144] https://books.google.co.uk/books/about/Clouds_and_Weather_Phenomena.html?id=iVMQCAAAQBAJ&redir_esc=y
[145] http://www.dailymail.co.uk/sciencetech/article-2160873/NASA-satellite-images-capture-commercial-jet-engine-trails-space.html
[146] http://www.checktheevidence.co.uk/cms/index.php?option=com_content&task=view&id=142&Itemid=50

References

[147] http://rapidfire.sci.gsfc.nasa.gov/subsets/?Europe_2_01/2007291/?Europe_2_01/2007290/Europe_2_01.2 007290.terra.1km.jpg
[148] http://rapidfire.sci.gsfc.nasa.gov/subsets/?Europe_2_01/2007291/Europe_2_01.2007291.terra.1km.jpg
[149] http://www.louthleader.co.uk/news/offbeat/circling-aircraft-revealed-to-be-from-nato-1-3425007
[150] http://rense.com/general81/ddthr.htm
[151] http://www.baha.be/Webpages/Navigator/News/tanker_flight_240205.htm
[152] http://www.rense.com/general81/detr.htm
[153] https://www.researchgate.net/publication/259885230_Formation_Properties_and_Climate_Effects_of_Contrails
[154] http://rapidfire.sci.gsfc.nasa.gov/subsets/?Europe_2_01/2007035/Europe_2_01.2007035.terra

[155] http://iopscience.iop.org/article/10.1088/1748-9326/11/8/084011/pdf
[156] https://www.sciencealert.com/first-published-study-on-chemtrails-finds-no-evidence-of-a-cover-up
[157] https://www.newscientist.com/article/2101611-chemtrails-conspiracy-theory-gets-put-to-the-ultimate-test/
[158] http://www.slate.com/blogs/bad_astronomy/2016/08/15/scientists_look_at_chemtrail_claims_and_find_them_lacking.html
[159] http://www.badastronomy.com/index.html
[160] http://www.smithsonianmag.com/science-nature/airplane-contrails-may-be-creating-accidental-geoengineering-180957561/
[161] http://journals.ametsoc.org/doi/abs/10.1175/1520-0442(2004)017%3C1123%3ARVIUDT%3E2.0.CO%3B2
[162] Gary Jones FOIA https://www.whatdotheyknow.com/request/contrails_chemtrails#incoming-284687
[163] This group was set up by Josefina Fraillle Martin in 2012 – http://www.guardacielos.org/?lang=EN
[164] Conference Report http://www.checktheevidence.co.uk/cms/index.php?option=com_content&task=view&id=370&Itemid=83
[165] Ms. Peterson was a Keynote Speaker at the 60th Annual DPI/NGO Conference on Climate Change (New York on September 5-7, 2007 http://www.agriculturedefensecoalition.org/content/about-rosalind-peterson
[166] Germany – https://www.youtube.com/watch?v=NT1xjMMAnEU
[167] USA – https://www.youtube.com/watch?v=NHRVmF8YkRc
[168] Bye Bye Blue Sky https://www.youtube.com/watch?v=dTxwDJ2ZDkk
[169] Aerosol Crimes – https://www.youtube.com/watch?v=dQuqAtVNnwY
[170] What in the World Are They Spraying? – https://www.youtube.com/watch?v=jf0khstYDLA
[171] http://www.checktheevidence.com/cms/index.php?option=com_content&task=view&id=370&Itemid=50
[172] https://www.prlog.org/12108727-beyond-theories-of-weather-modification.html
[173] https://www.facebook.com/mrmaxbliss?ref=br_rs
[174] http://www.geoengineeringwatch.org/
[175] http://www.geoengineeringwatch.org/geoengineering-methane-eruptions-and-imploding-arctic-ice/
[176] http://sonomachemtrails.blogspot.co.uk/2009/05/ac-griffin-talks-about-chemtrails.html
[177] http://www.geoengineeringwatch.org/ex-military-bio-environmental-engineer-kristen-meghan-blows-whistle-on-air-force-chemtrails/
[178] https://www.metabunk.org/threads/kristen-meghan-former-us-air-force-whistle-blower.1066/
[179] http://csat.au.af.mil/2025/volume3/vol3ch15.pdf
[180] http://planefinder.net/about/ads-b-how-planefinder-works/
[181] http://www.flightradar24.com/
[182] http://uk.flightaware.com/
[183] http://www.radarvirtuel.com/
[184] http://www.flightradar24.com/apps
[185] http://www.raspberrypi.org/help/faqs/
[186] http://www.satsignal.eu/raspberry-pi/dump1090.html
[187] See forum discussion: http://www.raspberrypi.org/forums/viewtopic.php?t=46063&p=364426
[188] http://www.checktheevidence.com/video/PiTrackerTL/2014-04-17-SN54HA-East-NE-10am%20onwards%20-%20trails%20smear%20out%20and%20form%20cloud%20.mp4
[189] http://www.checktheevidence.co.uk/cms/index.php?option=content&task=view&id=52
[190] http://www.checktheevidence.co.uk/cms/index.php?option=com_content&task=view&id=296&Itemid=50
[191] http://urlbam.com/ha/M002Z
[192] http://www.rense.com/general76/cars.htm
[193] https://www.amazon.co.uk/Goodnight-Mister-Longman-Literature-11-14/dp/0140315411/ref=sr_1_1?ie=UTF8&s=books&qid=1234818373&sr=1-1

194 http://www.spiritaero.com/about-spirit/
195 https://www.prisonplanet.com/global-chemtrail-bombardment-user-photos-and-video.html
196 http://www.mdpi.com/1660-4601/12/8/9375/pdf
197 http://www.sandiegouniontribune.com/sdut-journal-retraction-chemtrails-2015sep27-story.html
198 https://debunkingdenialism.com/2015/09/05/flawed-chemtrails-paper-by-herndon-retracted/
199 http://www.geoengineeringwatch.org/stephen-colbert-tells-david-keith-government-is-already-spraying-us/
200 http://members.tripod.com/~DELTA_9/index3.html
201 http://macedoniaonline.eu/content/view/6587/56/
202 https://www.accuweather.com/en/weather-blogs/weathermatrix/anomalies-radar-1/12444
203 http://www.checktheevidence.com/video/index.php?dir=PiTrackerTL/
204 http://www.baker-maker.com/2012/02/mackerel-sky.html
205 https://www.youtube.com/watch?v=wZYMCE3qbrE
206 http://weatherwars.info/2014/06/06/ten-years-ago/
207 http://chemtrails.foroactivo.com/t1400-30-de-enero-de-2009-desvian-borrasca-angulos-rectos
208 http://www.metoffice.gov.uk/satpics/latest_uk_ir.html
209 http://www.youtube.com/watch?v=rMfCO1U0jJw
210 http://weatherwars.info/?page_id=32
211 http://news.bbc.co.uk/1/hi/sci/tech/3394461.stm
212 http://earthsky.org/earth/andrew-heymsfield-on-hole-punch-clouds-made-by-jets
213 https://staff.ucar.edu/users/heyms1
214 https://www2.ucar.edu/about-us
215 http://journals.ametsoc.org/doi/pdf/10.1175/2009BAMS2905.1
216 http://weatherwars.info/?page_id=44
217 http://weatherwars.info/holes-part-2/
218 http://weatherwars.info/wp-content/uploads/2016/08/ – Search for "Sky Circle"
219 http://imageevent.com/firesat/strangedaysstrangeskies
220 http://www.dailymail.co.uk/sciencetech/article-1189877/The-cloud-Meteorologists-campaign-classify-unique-Asperatus-clouds-seen-world.html
221 https://www.youtube.com/watch?v=4rPKyr57v_E
222 https://youtu.be/HsclWbSKHlY
223 http://www.atoptics.co.uk/halo/dogfm.htm
224 https://www.jstor.org/stable/1520588?mag=charles-hatfield-rainmaker&seq=1
225 http://www.imdb.com/title/tt0049653/
226 https://daily.jstor.org/charles-hatfield-rainmaker/
227 http://www.latimes.com/local/california/la-me-rainmaker-20150526-story.html
228 http://www.nndb.com/people/776/000079539/
229 https://www.wired.com/2007/11/nov-13-1946-artificial-snow-falls-for-the-first-time/
230 http://langmuir.nmt.edu/Storms_Above/StormsAboveCh3.html
231 http://www.atmos.albany.edu/daes/bvonn/bvonnegut.html
232 http://just-clouds.com/hygroscopic_cloud_seeding.asp
233 http://www.weathermodification.com/projects.php?id=6
234 http://www.weathermodification.com/program-services.php
235 http://www.jfklibrary.org/Asset-Viewer/DOPIN64xJUGRKgdHJ9NfgQ.aspx
236 http://www.aoml.noaa.gov/hrd/hrd_sub/sfury.html
237 http://www.infowars.com/video/clips/weather_wars/wm_bb.htm
238 https://www.youtube.com/watch?v=MKl9rqw1Ykw
239 http://www.virtual.vietnam.ttu.edu/cgi-bin/starfetch.exe?xcPaQZKJOuE0cy2V0fFih3Fj44V3ddY0rf6SOuxl@8KQPA04mOs.DbzZxzVWIgdo.7xc2hHR9wuuZrTtKTBoFw.V8b6@zNuD@LggGbLmz3s/2390601002C.pdf
240 http://www.opsecnews.com/operation-popeye-weaponized-weather-during-vietnam-war/
241 http://archive.defense.gov/Transcripts/Transcript.aspx?TranscriptID=674
242 https://www.metabunk.org/debunked-others-are-engaging-even-in-an-eco-type-of-terrorism.t159/
243 http://csat.au.af.mil/2025/volume3/vol3ch15.pdf
244 http://www.gi.alaska.edu/haarp/faq
245 http://www.gi.alaska.edu/alaska-science-forum/haarp-again-open-business
246 http://www.haarp.net/
247 https://fas.org/irp/program/collect/haarp.htm
248 https://ahrcanum.com/2009/10/16/earthquake-haarp-connection-evidence/
249 http://patft.uspto.gov/netacgi/nph-Parser?Sect1=PTO1&Sect2=HITOFF&d=PALL&p=1&u=/netahtml/PTO/srchnum.htm&r=1&f=G&l=50&s1=4,686,605.PN.&OS=PN/4,686,605&RS=PN/4,686,605
250 http://www.eiscat.se/about/whatiseiscat
251 http://www.arrl.org/news/haarp-like-ionospheric-research-project-underway-at-arecibo-observatory
252 http://www.abc.net.au/local/audio/2011/04/05/3182616.htm

References

[253] http://www.nature.com/news/2009/091002/full/news.2009.975.html
[254] http://www.redicecreations.com/radio/2011/01/RIR-110125.php
[255] http://www.pfrr.alaska.edu/
[256] http://www.swpc.noaa.gov/sites/default/files/images/u33/McCoy_SWW2016.pdf
[257] http://www.viewzone.com/haarp44.html
[258] https://www.amazon.co.uk/Planet-Earth-Dr-Rosalie-Bertell/dp/1551641828
[259] http://www.wilhelmreichtrust.org/books.html
[260] http://www.checktheevidence.com/cms/index.php?option=com_content&task=view&id=393&Itemid=50
[261] https://www.youtube.com/watch?v=pllRW9wETzw
[262] https://youtu.be/DJ29VcZeNDg?t=2m00s
[263] https://youtu.be/Cv8PjGbtufU?t=2m45s
[264] https://www.amazon.co.uk/d/cka/Book-Dreams-Peter-Reich/1784182702
[265] https://youtu.be/3B4lIK9aS_E?t=5m9s
[266] http://www.wilhelmreichtrust.org/biography.html
[267] http://www.orgonomy.org/
[268] http://www.wilhelmreichtrust.org/
[269] http://www.wilhelmreichtrust.org/trust_history.html
[270] https://www.researchgate.net/publication/233954668_EXPERIMENTAL_CONFIRMATION_OF_THE_REICH_ORGONE_ACCUMULATOR_THERMAL_ANOMALY
[271] https://www.abebooks.co.uk/9780682478229/Planet-Trouble-Ufo-Assault-Earth-0682478229/plp
[272] http://educate-yourself.org/ct/goodbyects10jan02.shtml
[273] http://www.orgonelab.org/ResearchSummary2.htm
[274] http://www.keelynet.com/news/042714i.html
[275] https://www.youtube.com/watch?v=YPXj65d6mW8
[276] https://www.youtube.com/watch?v=eBck494b4Pg
[277] http://www.rainengineering.com/
[278] http://www.drjudywood.com/articles/NIST/Qui_Tam_Wood.shtml
[279] http://tinyurl.com/9/11ftb
[280] https://www.youtube.com/results?search_query=dr+judy+wood+where+did+the+towers+go
[281] http://www.drjudywood.com/articles/a/bio/Wood_Bio.html
[282] http://www.drjudywood.com/
[283] http://www.drjudywood.com/articles/erin/
[284] http://9/11digitalarchive.org/REPOSITORY/IMAGES/PHOTOS/1867.pjpeg
[285] http://www.youtube.com/watch?v=WDVYlDaXsFg
[286] http://cimss.ssec.wisc.edu/tropic/archive/2001/storms/erin/erin.track.gif
[287] http://tropic.ssec.wisc.edu/storm_archive/2001/storms/erin/erin.html
[288] http://www.drjudywood.com/articles/erin/erin5.html
[289] http://www.thehutchisoneffect.com
[290] https://www.dailykos.com/stories/2005/9/7/146582/-
[291] https://grahamhancock.com/kreisbergg2/
[292] https://www.reddit.com/r/askscience/comments/jufux/can_somebody_explain_the_em_lines_on_this_recent/c2f926c/
[293] https://www.theguardian.com/commentisfree/2017/aug/28/climate-change-hurricane-harvey-more-deadly
[294] https://sustainabledevelopment.un.org/content/documents/Agenda21.pdf
[295] http://fortune.com/2015/02/25/they-tried-to-arrest-me-for-planting-carrots/
[296] http://abcnews.go.com/US/vegetable-garden-brings-criminal-charges-oak-park-michigan/story?id=14047214
[297] https://www.biggreenradicals.com/group/greenpeace/
[298] http://www.postsustainabilityinstitute.org/board-of-directors.html
[299] http://www.britannica.com/topic/communitarianism
[300] http://www.rand.org/topics/delphi-method.html
[301] http://fellowshipoftheminds.com/2012/05/23/agenda-21-huffpo-hacks-bias-exposed-by-rosa-koire/
[302] http://www.huffingtonpost.com/andrew-reinbach/agenda-21-sustainability-_b_1523118.html
[303] http://www.endagenda21.com/
[304] http://www.iclei.org/
[305] http://www.iclei.org/details/article/rockefeller-foundation.html
[306] http://www.cpexposed.com/
[307] http://commonpurpose.org/knowledge-hub/all-articles/smart-cities-cities-as-magnets/
[308] http://www.justice.gov/opa/pr/launch-strong-cities-network-strengthen-community-resilience-against-violent-extremism

309 http://strongcitiesnetwork.org/
310 http://www.theguardian.com/local-government-network/2014/mar/21/city-mayors-cities-green-new-york-copenhagen
311 http://www.nyc.gov/html/ia/html/about/un.shtml
312 https://www.lucistrust.org/about_us/history
313 https://youtu.be/1hIr-ZlqIko?t=6m43s
314 http://www.freeminds.org/history/cemetary.htm
315 https://www.lucistrust.org/the_great_invocation/eleanor_roosevelt_reads_the_great_invocation
316 https://www.lucistrust.org/the_great_invocation
317 https://www.lucistrust.org/blog/blog_world_goodwill_at_the_un/entry/sustainable_development_goals_the_united_nations_and_the_libra_full_moon
318 https://www.lucistrust.org/conferences/show/conference_london_2016
319 https://youtu.be/dh-gT2jsIrc
320 http://www.bbc.co.uk/news/uk-scotland-south-scotland-35014930
321 http://www.telegraph.co.uk/news/uknews/1556544/Hull-is-forgotten-city-of-the-flooding-crisis.html
322 http://www.visitcumbria.com/cockermouth-floods/
323 http://www.bbc.co.uk/news/uk-20488645 http:/www.wired.co.uk/news/archive/2015-10/20/met-office-storm-names-barney-steve-wendy
324 http://www.telegraph.co.uk/comment/columnists/christopherbooker/10625663/Flooding-Somerset-Levels-disaster-is-being-driven-by-EU-policy.html
325 http://ec.europa.eu/environment/nature/natura2000/index_en.htm
326 http://www.manchestereveningnews.co.uk/news/greater-manchester-news/policing-barton-moss-fracking-site-6473574
327 http://www.talkfracking.org/news/nana-threatened-with-arrest-for-offering-police-tea-and-cakes/
328 http://www.cop21paris.org/about/cop21
329 http://www.theguardian.com/world/live/2015/nov/13/shootings-reported-in-eastern-paris-live
330 https://www.youtube.com/watch?v=vKJWZaZL1ME
331 https://www.youtube.com/watch?v=JKvkhe3rqtc
332 http://www.independent.co.uk/news/world/europe/global-climate-change-2015-police-fire-tear-gas-into-crowds-as-100-protesters-arrested-in-paris-a6753621.html
333 https://www.theengineer.co.uk/issues/9-april-2007/stephen-salter-pioneer-of-wave-power/
334 http://www.youtube.com/watch?v=_bdeNuRF-yE
335 http://www.vestaldesign.com/blog/2006/09/conspiracy-salters-duck/
336 http://www.top-alternative-energy-sources.com/stanley-meyer.html
337 http://www.panacea-bocaf.org/unitednuclearhydrogencars.htm
338 https://www.youtube.com/watch?v=K3GDjVskYIs
339 https://www.youtube.com/watch?v=dIz29JHJmMk
340 http://www.youtube.com/watch?v=CkVjONutBQ0
341 http://freedomvideo.org/2010/05/fire-from-water/
342 http://www.infinite-energy.com/whoarewe/gene.html
343 http://www.thmoray.org/
344 http://waterpoweredcar.com/stanmeyer.html
345 https://www.bibliotecapleyades.net/disclosure/briefing/disclosure13.htm
346 http://tinyurl.com/9/11wrh
347 http://www.checktheevidence.com/cms/index.php?option=com_content&task=view&id=406&Itemid=50
348 https://www.youtube.com/user/dutchsinse/search?query=erin
349 https://www.youtube.com/watch?v=bbOVdi4mLrs
350 https://youtu.be/FeGpS0mnHt4

Printed in Great Britain
by Amazon